Praise for

THE GOOD ASSASSIN

"What a wonderful book. Stephan Talty's fast-paced account of how Herbert Cukurs, the Latvian aviator turned Nazi war criminal, was eventually brought to justice by Mossad operatives is as gripping as any novel. Hard as it is to read the details of Cukurs' horrific crimes, the outcome is both moving and uplifting, with the Latvian's demise helping to bring other perpetrators of genocide to justice. Talty is at the top of his game."

— SAUL DAVID, AUTHOR OF *OPERATION THUNDERBOLT* AND *THE FORCE*

"Part Holocaust history, part detective case, part spy operation, *The Good Assassin* is an enthralling book. Stephan Talty paints vivid, often chilling portraits of its vengeful hero, Mossad agent Jacob Medad, and Herbert Cukurs, the war criminal he pursued to the bitter end. It's a stunning, you-are-there kind of read."

— NEAL BASCOMB, *NEW YORK TIMES* BEST-SELLING AUTHOR OF *HUNTING EICHMANN* AND *FASTER*

"Stephan Talty's *The Good Assassin* is a gripping chronicle of one of the most brilliant operations launched against an escaped Nazi war criminal, and a fitting memorial to the victims of the Holocaust in Latvia and to the brave Israelis who traveled halfway around the world to punish one of the key perpetrators of those crimes. At a time when Latvian ultranationalists are trying to rehabilitate Cukurs as a national hero, Talty explains why such a step would be a grave miscarriage of justice."

— DR. EFRAIM ZUROFF, CHIEF NAZI-HUNTER, SIMON WIESENTHAL CENTER

"Stephan Talty masterfully recounts how the Holocaust engulfed the Jews of Latvia and how the architect of that genocide was hunted to his death by Israeli spies. It's a page-turning account of a little-known episode of the Shoah and how justice was brought to one of its key perpetrators."

— PETER BERGEN, AUTHOR OF
MANHUNT: THE TEN-YEAR SEARCH FOR BIN LADEN AND
TRUMP AND HIS GENERALS: THE COST OF CHAOS

"Stephan Talty has crafted a fast-paced account of an overlooked part of the Holocaust—and its broader impact on the postwar hunt for its perpetrators."

— WILLIAM GEROUX, AUTHOR OF
THE GHOST SHIPS OF ARCHANGEL AND *THE MATHEWS MEN*

"Talty efficiently mines archival records for vivid details and tracks the complexities of Medad's undercover mission with flair. The result is a captivating and gruesome real-life spy thriller."

— *PUBLISHERS WEEKLY*

"Compelling . . . Talty remains true to his technique, delivering thoroughly researched, engrossing nonfiction in a thrillerlike narrative style . . . As anti-Semitism surges once again, this page-turning history reminds us of the sanguinary consequences of unchecked hatred."

— *KIRKUS REVIEWS*

"Thrilling . . . A fast-paced, recommended work that enthralls, edifies, and reveals the disturbing extent to which Latvians and others participated in genocide."

— *LIBRARY JOURNAL*

"The author brings his usual attention to detail, excellent research, terrific storytelling, passion, and dedication to this suspenseful recounting of a shadowy facet of the Holocaust, which continues to haunt the world."

— *BOOKLIST*

THE GOOD ASSASSIN

BOOKS BY STEPHAN TALTY

NONFICTION

Saving Bravo

The Black Hand

Agent Garbo

Escape from the Land of Snows

The Illustrious Dead

Empire of Blue Water

Mulatto America

FICTION

Hangman

Black Irish

THE GOOD ASSASSIN

How a Mossad Agent and a Band of Survivors Hunted Down the Butcher of Latvia

STEPHAN TALTY

Mariner Books
Houghton Mifflin Harcourt
Boston New York

First Mariner Books edition 2021

hmhbooks.com

Library of Congress Cataloging-in-Publication Data
Names: Talty, Stephan, author.
Title: The good assassin : how a Mossad agent and a band of survivors hunted down the Butcher of Latvia / Stephan Talty.
Other titles: How a Mossad agent and a band of survivors hunted down the Butcher of Latvia
Description: Boston ; New York : Houghton Mifflin Harcourt, 2020. | Includes bibliographical references and index.
Identifiers: LCCN 2019027262 (print) | LCCN 2019027263 (ebook) | ISBN 9781328613080 (hardcover) | ISBN 9781328618856 (ebook) | ISBN 9780358522478 (pbk.)
Subjects: LCSH: Medad, Jacob, 1919–2012. | Cukurs, Herbert, 1900–1965 — Assassination. | Israel. Mosad le-modiʻin ṿe-tafḳidim meyuḥadim. | Nazis — Brazil. | Nazis — Uruguay. | War criminals — Latvia — Rīga. | Fugitives from justice — Uruguay. | Jews — Persecutions — Latvia — Rīga. | Holocaust, Jewish (1939–1945) — Latvia — Rīga. | Secret service — Israel.
Classification: LCC D804.L35 T35 2020 (print) | LCC D804.L35 (ebook) | DDC 364.15/1092—dc23
LC record available at https://lccn.loc.gov/2019027262
LC ebook record available at https://lccn.loc.gov/2019027263

Book design by Emily Snyder

Printed in the United States of America
DOC 10 9 8 7 6 5 4 3 2 1

For Karen, Suraiya, and Aman

CONTENTS

PROLOGUE

The Apartment on the Avenue de Versailles

MIO WALKED INTO THE LOBBY of the building on the Avenue de Versailles and called out *"Bonjour!"* to the concierge through her tiny window. Not waiting for a response, he went quickly up the marble stairs. Puffing by now — he was a bit out of shape — he reached the wooden door of Yariv's apartment and pressed the bell. He was confident he hadn't been followed; he'd stopped in front of the gigantic Radio France building a few blocks away to check for tails. It would have been unfortunate to bring one to a meeting with Yariv, who was touchy about such things.

The door opened and Yosef Yariv, the head of Caesarea, the special operations arm of Mossad, nodded at Mio. With his honking beak of a nose and thick pelt of unruly hair, the forty-year-old Yariv resembled a predatory desert bird. Now his piercing blue-gray eyes studied his friend.

"I'm glad you made it," he said.

Mio said nothing, only nodded and walked past. Yariv locked the door, then turned. "From this moment onwards," he said, "your name is Anton Kuenzle. You'd better start getting used to it." Mio showed no reaction; he was an introvert, raised in Germany as a Jew in the early thirties, which encouraged, if not required, certain kinds of masks to be worn. And besides, it was Mio's stock-in-trade to become different people, sometimes for a few days, other times for much longer. Inside Mossad, where he was one of the great, perhaps the greatest, undercover operatives, he was known as "the man with the hundred identities." Back home in Israel, his family lived in a house that sat behind a steel gate, through which the agency sent a car every time he was leaving on an assignment. His son would later say that when the car drove off and they heard the loud clang as the gate swung closed behind it, they knew their father had already transformed into another person. Calling it a cover identity wasn't quite right; when Mio assumed a new persona, it didn't cover anything, let alone his real personality. It *was* his real personality, for exactly the length of time he was required to inhabit it. A fellow Mossad agent once claimed, "I swear to God," if you woke Mio in the middle of the night, he would immediately begin speaking in the language of his false persona. On those days when he was driven to the airport, he never looked back to wave to his children because, in his mind, he had no children.

The two walked ahead into a small guest room. Another operative — Mio called him Michael, though that wasn't his real name — sat at a small table with cups and saucers and a pot filled with coffee. A "fairly thin" file sat next to the cups. Mio nodded at Michael and took one of the empty chairs. Yariv followed suit. He looked at the other two, his eyes cool.

"You must be wondering why I summoned you here," he said.

The two men said nothing.

"Well, it all begins with the final confirmation we received about a Nazi war criminal who lives in one of the South American countries."

Michael looked at Mio, who glanced back, remaining silent. Yariv explained that in eight months, on May 8, 1965, the world would mark the twentieth anniversary of the end of World War II in Europe. German politicians and ordinary citizens were calling for an end to the hunt for Nazi war criminals and for a statute of limitations to be applied to their crimes. Mio didn't react, but, as someone who avidly read the newspapers, he must have seen the headlines. Germany was preparing to enforce an 1871 law that mandated a twenty-year limit on murder prosecutions. Two other amnesties, for assault and for manslaughter, had gone into effect in 1955 and 1960 with little protest around the world. Charging any Nazi officer or soldier with those crimes was now forbidden inside Germany. But soon the killers themselves, the very worst of the worst, the men and women who'd physically pulled the triggers on the machine guns and the rifles and the pistols and smashed in the heads and strangled and bludgeoned their portion of the six million, could emerge from their hiding places and walk free in the sun. It seemed utterly fantastic, but there it was.

The statute, Yariv said, was popular in West Germany. Every poll showed solid majorities in favor of it, and the governing party, the Christian Democratic Union, had thrown its weight behind the law. Only the Bundestag, the feisty German parliament, could delay the amnesty by passing a bill that would push the deadline a few years into the future, allowing the remaining unindicted National Socialist murderers to be found and prosecuted for at least a short time longer. But Yariv told Mio and Michael that Israeli leaders were increasingly pessimistic about this possibility. "The chances of accepting this proposal are small . . . There is no guarantee that the politicians are prepared to extend the Statute of Limitations, not by four years, not by ten years, and, for that matter, most probably not at all."

Mio noticed his friend's voice starting to rise in the quiet room, though his face showed no change in expression. "It is absolutely inconceivable," Yariv said, "that tens of thousands of Nazi war criminals, who never paid for their heinous crimes, should now be able to crawl out of their hiding holes and spend the rest of their lives in peace and tranquility . . . It's been only twenty years since the release of the survivors of the death camps, and we owe it to them, and to the six million who did not survive and are unable to avenge themselves — we must thwart this shameful process."

Israeli prime minister Levi Eshkol and his intelligence chiefs had secretly decided on a mission. A killing was required, a certain kind of killing that would reveal the Nazi monsters who'd escaped punishment and publicize the nature of their crimes. Unlike Mossad's kidnapping and subsequent execution of Adolf Eichmann four years earlier, there would be no trial, no lawyers or judges, no legal niceties, no essays by Hannah Arendt in *The New Yorker.* And the operation had to be completed before the vote in the German parliament, currently scheduled for sometime in the spring.

"The Nazi whose turn has come," Yariv said, "is Herbert Cukurs."

It was a Latvian name; Yariv pronounced the "C" in "Cukurs" correctly, like "Ts," *TSOO-krz.* (It means "sugar.") At a conference of Israeli intelligence chiefs in January, the names of potential assassination targets had been read out. When the speaker came to Cukurs, one of the men in the room collapsed. It was Major General Aharon Yariv—no relation to Yosef Yariv—head of the country's Military Intelligence Directorate. Cukurs had murdered several of Yariv's loved ones and friends during the war; his reaction was one reason why the Latvian's name had been chosen.

Mio had never heard of Cukurs, and he showed no emotion at the idea of ending his life. "Outwardly," he said, "I kept a poker face." If he was chilly in his personal life—and he was, to his children's eternal regret—he was even more clinical when working. A quickening of the breath, a raised eyebrow, would for him have been a breach of professional ethics. But inside, he was deeply stirred. His mother and father, a German patriot and a recipient of the Iron Cross for bravery in World War I, who'd believed that they'd be saved until almost the very end, had been murdered at Auschwitz and the "model" camp of Theresienstadt. Despite his outward calm, when Mio heard the Nazi's name, he said, "I felt my heart and my adrenaline level skyrocket suddenly."

"We are not dealing here with a desk murderer like Eichmann," Yariv went on. "[Cukurs] is personally responsible for the annihilation of at least 30,000 Jews in Riga." And unlike more famous men like Dr. Josef Mengele, whom Mossad had been unable to find despite two decades of searching, Cukurs' whereabouts had been confirmed. He was living in a small house in São Paulo surrounded by guard dogs and a barbed wire fence. Yariv looked at Mio. "I propose that you . . . go to Brazil disguised as an Austrian businessman under the name of Anton Kuenzle." Under this light cover, he would find the Nazi, befriend him, infiltrate his circle, and arrange his death. The execution would then be announced to the world, and (Mossad hoped) the news stories about the savage killer, his grateful victims, and his faceless assassins — forced to act as the authorities in Berlin and other European capitals dawdled — might just convince the Germans that going ahead with the amnesty was an impossibility. "I'm well aware that this is no simple task," Yariv said. "You will face a criminal who is, according to our reports, cunning, mistrustful, ruthless and dangerous, and is always prepared for the worst."

Michael began leafing through the file that sat on the table. "Will Mio actually operate alone," he said, "or will we send a small stalking and protection unit with him?"

For the first time that morning, Mio spoke up. "I prefer to work alone," he said. "Me against the target."

Yariv nodded, then gestured to the file, a handful of pages in a manila folder that barely rose above the lacquered surface of the table. The slimness of the file was significant for reasons that only those intimate with the history of the Latvian Shoah would understand. It was so thin because so few Jews had been left alive to speak about Herbert Cukurs. Inside the folder were perhaps half a dozen testimonies — the exact number isn't known — that traced Cukurs' actions during the war, painstakingly collected from eyewitnesses living in several countries during the late 1940s and early 1950s. Some of the accounts were barbaric, others oddly moving. In one story, Cukurs speaks to a young girl in Yiddish; they have a short, pleasant conversation before the Latvian, for no apparent reason, pulls out his Russian handgun and executes her in cold blood. In another, he saves a woman he knew to be Jewish, at considerable risk to his own life. The collection, in fact, added up to a curiously fractal, incomplete portrait of Herbert Cukurs, whose life had been larger and stranger than

Mio could imagine at that moment; it would take many years and the survival of one obsessed young Jewish woman to tell it in full. "Overall, I must say he is a fascinating historical figure," one survivor later wrote, "full of tremendous contradictions." Though they could not have known it that morning in Paris, the Israelis had chosen for elimination a symbol of the Shoah whose life would speak to the motivations of those accomplices in eastern Europe who had carried it out.

With the preliminary details settled, Mio and Michael began sorting through the pages; they each picked up a selection and began to read. The white china cups, the husky September light streaming through the window, a blurred car horn from the street below, the loveliness of a fall afternoon in the Sixteenth Arrondissement faded from their thoughts as the testimonies inside ushered them to the city of Riga in the black year of 1939.

PART ONE

The *Aktions*

ONE

The Club on Skolas Street

THE DOUBLE WOODEN DOORS of the building at 6 Skolas Street, tucked into a thick archway topped by garlands cut into white stone, stood open, and the lights inside glowed as the crowd streamed in out of the blade-sharp December cold, the men dressed in fedoras and dark suits with generously cut lapels, white shirts, and dark ties, the women in simple hand-sewn dresses or Paris frocks, a few in fur coats, glossy and rich to the touch. The building, Riga's Jewish Club, was well known to almost everyone in attendance that winter night. Who among them hadn't come here before to see a play on a crisp fall evening, or to sink into the music of a quartet playing Schubert and forget the fast-arriving December dusks? Who hadn't been dragged here for a second cousin's wedding, where the sound of the violin bounced off the marble floor and it was wall-to-wall aunts and uncles wherever you looked? Or who hadn't simply come, on a stifling Sunday afternoon during the summer, to take a Yiddish novel down from the library shelves and quietly while away the afternoon?

But that night in 1939, the crowd wasn't there for another Orthodox wedding or a local rabbi's speech on the meaning of Purim. It was a rather special night. A world-renowned Latvian aviator, in fact the only world-renowned Latvian aviator, had come to talk to the audience members about Palestine—and not just the idea of Palestine, which was something Riga's Jews had debated endlessly, to the point of grim exhaustion, but the place itself, its odors and its aquifers, its sunburned pioneers. Riga was home to many, often feuding persuasions of Jews: there were the Zionists, the revisionist Zionists, the socialist Zionists, the Bundists, the followers of the Mizrachi Party, the ultra-Orthodox, and the communists, to name just a few. Representatives of most of the factions were at 6 Skolas Street that night; a sort of truce prevailed. "The entire Jewish elite gathered," said one young man in attendance. The black-bordered advertisements that ran in Riga's newspapers the week before had featured the speaker's name prominently: Herbert Cukurs.

Almost anyone who lived in Riga and was even marginally acquainted with what was going on in the world would have been familiar with the name. Cukurs was known as "the Latvian Lindbergh," and though some Latvians, as occupants of a small country with an often tragic history, tended to exaggerate the accomplishments of their native sons, his wasn't such a case. As famous as Lindbergh was in America, Cukurs was equally lionized in Latvia and in certain other quarters of Europe. His accomplishments were as solid and familiar as the stone cladding of the building the Jews stood in.

The audience members made their way into the auditorium and took their seats, voices thrumming against the paneled walls topped by wooden cutouts of Greek urns overflowing with vines. Up above, latecomers flowed into the balcony. Hundreds of tickets had been sold for the night's presentation, and the hall was approaching capacity. There was a stage at the far end and a wooden lectern with a reading board tilted upward; behind it sat a large white screen. The seats slanted down toward the lectern, affording everyone a decent view. The eyes of the Jewish men and women focused on the broad-shouldered thirty-nine-year-old aviator sitting on the stage, his pomaded dark brown hair flashing under the lights when he turned his head to talk to the men on either side of him.

At the appointed time, the aviator was introduced; he stood and strode toward the lectern. Among the pale, shy Latvians, often described as the introverts of Europe, here was a man who practically beamed out rays of physical confidence. He was thick-chested, robust through the legs, and crudely handsome, with a broad face that narrowed toward the chin, a Roman emperor's nose dipping hawkishly above the lips, and intelligent blue-gray eyes below dark eyebrows. His gleaming hair was cut short at the sides, longer on top, with a part on the center-left, in the dashing style of Clark Gable. There was nothing fine-boned or melancholy about Herbert Cukurs; he looked thoughtful but tough. He was both of those things.

The aviator may have been the most admired man in the country at that moment, but many audience members understood he wasn't a simple proposition vis-à-vis the Jews. Cukurs was known as a nationalist to the bone, and among his friends and associates, dark comments about Yiddish speakers were almost de rigueur. "As was the custom among military circles, he also made the occasional anti-Semitic remark," wrote one Jew many years later. "But this was nothing extraordinary." Anti-Semitic humor was in the air in Riga, as it was everywhere in eastern Europe. What could you do? "He was not really considered a Jew-hater," the man continued. "The opposite was true, since he was often seen in Riga's cafes in the company of Jewish intellectuals."

The famous pilot had grown up in the western city of Liepāja in a Christian farming family with four sons, whose father had eventually sold his acreage, moved to town, and opened a machine shop. The family worked hard and treated their employees fairly, including the Jewish apprentices whom they often hired. "Jānis Cukurs and his wife were decent, honest folks," said one of the apprentices, Meir Deutsch. "They even had a warm attitude toward the Jews." At the time, Jews were restricted from working in certain jobs: at the universities, in Latvian-owned factories, in city administration, and on the police force (which hired only one token Jew). As employers, the Cukurses were fairly liberal for the era.

Herbert was always different from his three brothers, and always *wished* to be different. He was naturally rebellious, following orders with difficulty or not at all. He was also, in the local parlance, "sky-crazy" — that is, infatuated with flying. At age twenty, the strapping, athletic Cukurs joined the struggling Air Force of the newly independent Latvia, when all the pilots had to command were antique, worn-out German and British planes that "would not take off until the last dust was blown off the engines." Aviation was frequently a lethal business in 1920, and Latvia's brutal winters made it even more so, as the frigid air tended to freeze oil lines and motors. (In December and January, the locals say, "the frost bites so hard that a hare's eyes jump out of their sockets.") Many times, Cukurs would have to wrestle a plane with a sputtering engine down from the sky and land it, at least once on the frozen surface of Stropi Lake.

Despite the risk, the "young hothead from Liepāja" grew fond of dreaming up wild stunts. He would take to the skies in the aircraft he'd built himself out of scrap parts and perform barrel rolls, zooming upside down over Riga or Liepāja to impress the local girls. His plane "spun like a leaf released into the wind, then threw itself downwards like an arrow above the heads of the spectators." When he wasn't flying, Cukurs was fond of parading through the streets of Liepāja with his plane, its wings removed to fit through the narrow lanes. (A town official finally put a stop to this.) But he was also very good at what he did. While he was in the Latvian Air Force, the young Cukurs would sweep over the local fields, practicing bombing runs that required him to drop dummy explosives as close to the wooden targets as possible. The results were recorded by an Air Force photographer for future study. It was said that after all the pilots had completed their runs, the more senior officers in the squadron would quietly sidle up to the cameraman and ask him if he might "correct" the results, as Cukurs' bombs always seemed to land closer to the targets than theirs.

Cukurs was a dreamer, impetuous, physically fearless, difficult to control, hardworking, and fantastically ambitious. He seemed to feel that Latvia, which he loved, was too small for him; he'd been born for the world. In 1924, he left the Air Force, some said because of problems with his eyesight, while others reported he was forced to depart after a stunt where he flew under a local bridge, leaving only three feet between his tires and the surface of the water. He took a job as a truck driver and began designing what he hoped would be a revolutionary one-seater plane. With no money to spare for new parts, he picked flight instruments from scrap dealers' bins and restored them as best he could. With his young wife, Milda, he spread lacquer over the wooden planks, hammered them together, glued the wing ribs, and machined the fittings. When an Air Force aviator or private pilot crashed around Riga, Cukurs would race to the scene, pluck the canvas off the small nails that attached it to the aircraft's frame, and bundle it into his truck. He would bring the canvas home, cut it to size, and spread it across the skeletal fuselage of the plane he had deemed the C3 (the "C," of course, for Cukurs). His dream plane slowly took shape over years of backbreaking work. To earn money, he moonlighted on construction sites; "with bloody hands [he] lifted the big sharp fragments of blown-up stones into the truck, rolled cement barrels and loaded rails from sunrise to sunset." There were rumors that Cukurs occasionally earned even more cash "in a not entirely legitimate way" — by smuggling cases of illegal liquor.

In 1933, two Latvian pilots, Nikolajs Pulins and Rudolf Celms, announced they were going to fly the 4,700 miles from Riga to The Gambia, the tiny African nation tucked into the coast of Senegal in West Africa. The Gambia had played a part in national lore since Duke Jacob of Courland, whose lands later formed part of Latvia, established a mercantile colony there during the seventeenth century. For a brief time, the duke and his subjects boasted their own miniature African empire. The colony had long since disappeared, taken over by the British, but a vestigial memory of its existence remained lodged in the Latvian imagination. As Germany had ruled over what would later become Namibia and the USSR claimed its many satellites, so Latvia—tiny, insignificant Latvia!—had once been daring enough to plant a flag below the equator. So when the pilots announced that they were going to fly to The Gambia, it caused "thrills and expectations among the people."

But the next month, bad news. The explorers' plane had crashed in Germany, only a few days into the journey. The mission was over.

In stepped Cukurs. The thirty-three-year-old aviator announced he would pick up the gauntlet and fly to The Gambia, hopscotching south and west aboard the open-cockpit aircraft he'd built out of cast-off and salvaged parts. He would even name the aircraft *The Duchess of Courland*, in honor of the former ruler. The young flier hurried to finish his plane, which lacked even that most basic component: an engine. He scoured the countryside for a sturdy motor he could afford and eventually found an outmoded 80-horsepower Renault, built in 1916, that was available for scrap prices. It wasn't exactly a state-of-the-art power plant; this particular model was so old that the only other one of its kind was sitting in a French aviation museum. Cukurs purchased the antique and winched it into the engine compartment, then bolted it to its moorings. He was ready to take flight.

Almost nobody, however, cared. The airfield he took off from was practically deserted; only Cukurs' wife and son, the head of the Latvian Aviation Union, and an anonymous bystander saw *The Duchess of Courland* lift into the sky on the afternoon of August 28, 1933. The pilot, dressed in warm flying gear and wearing goggles, gained altitude, waved to the tiny figures from his open cockpit, then turned the nose of the plane south.

The journey began badly. Cukurs nearly collided with a plane in a fog bank above Paris, then came under attack from a French antiaircraft battery near Poitiers. (He had accidentally flown over a practice range for artillerymen, who lobbed explosive shells toward his cockpit.) French aviators were so impressed by his skill and daring that after he landed, they dressed him in one of their flight suits and carried him on their shoulders to a nearby bar to get drunk. Over Barcelona, one of the plane's cylinders failed, then another. The plane dove toward the sea, "the engine jumping and jerking like crazy, the frame seemingly broken." Cukurs spotted a dry riverbed and pushed the nose toward it. The wheels caught on impact and the plane flipped over, trapping Cukurs beneath the fuselage as the fumes from leaking oil slowly overcame him. He lay there for some time, until a soldier shook him awake and pulled him out from underneath the *Duchess.* In the sky again over Málaga, the propeller froze, and the plane lost power. Cukurs glided between two crags and put the plane down in a garden.

But these were trifles compared with what faced him when he reached Africa. As he flew over Casablanca, Arab tribesmen fired their rifles at his plane, and the wind spun the ocean into a froth, "everything boiling like a pot of witches." As he crossed the Sahara, the oil pressure gauge dropped to zero, and the Renault engine struggled to keep its cylinders from seizing up. When he landed at Dakar, the last stop before his final destination, the British consul informed him that, regrettably, there was no airfield in all of The Gambia for him to land on; if he insisted on flying there, he would most likely end up running out of gas and crashing into the jungle, where he would certainly be eaten by cannibals. A French official happened by and informed Cukurs that he'd heard there was, in fact, an aerodrome somewhere near the Gambian capital of Bathurst. What to do?

"Let's take the risk!" Cukurs cried. He gripped the propeller of the *Duchess,* spun it until the engine sputtered to life, then took off.

Hours later, as Cukurs searched the leafy canopy beneath him, the mouth of the Gambia River came swimming into view, then the modest capital, which he circled three times as he looked for a landing spot. Nothing. Only swamps and curling tributaries, the water glinting through the dark canopy. Cukurs crossed back and forth but saw not a single place where he could set the plane down. Finally, after venturing eight miles out from the capital, he spotted an aerodrome; low on gas, he settled *The Duchess of Courland* down on the airstrip.

The news of Cukurs' arrival electrified capitals from Riga to Rome; it was the kind of between-the-wars adventure that made a person proud to be a European. "What has been held impossible has been achieved!" crowed the head of the Latvian Aviation Union. The hothead from Liepāja had put his country on the world map, and Latvians were over the moon. Donations poured in for a new engine to replace the dilapidated old Renault that had powered the *Duchess* to Africa. It was even said that every Latvian mother now considered Cukurs to be an honorary son.

Cukurs knew nothing of this at first. He was sleeping on the streets in Bathurst, hoping that the reports he'd heard of man-eating leopards were unfounded and experiencing the first debilitating symptoms of malaria. Nonetheless, he was having a marvelous time. Wherever he went, Cukurs projected a devil-may-care insouciance that earned him the admiration of journalists and government officials alike. When he landed in Japan on a later journey, he "commanded great applause and . . . began making himself delightful to all around him. 'Oh, the object of my flight?' he quipped to a reporter. 'Nothing particularly profound. Attracted by the mystery of the East, I wish to wander around the strange regions I longed to see for so long.'" During another trip, as he was about to set off on a tiger hunt, a correspondent asked if he thought the sport dangerous. "Of course, the tiger may eat me," Cukurs replied. "I wish him *bon-appetit.*"

The British, who controlled The Gambia, were less than thrilled by the arrival of this charming interloper. Perhaps they were put off by Cukurs' unusual habit of treating the native people like human beings. Despite being told that if he ventured outside the main cities, he'd be killed by headhunters, the flier made sure to approach the locals with respect. "I'm winning the sympathy of the black people," he wrote, "as I'm probably the first European to treat them as equals, to address them as 'Mister' instead of whistling at them." This sentiment seemed to strike a chord with the two million Latvians, whose country had been brutally raked over by wars and occupied by Germany and Russia in turn before gaining its independence in 1918. Latvians knew what it meant to be colonized through terror.

Cukurs flew back to Latvia, enduring typhoons, malaria, and the incessant chattering of two young monkeys, Gao and Timbuktu, a gift from French aviators that he kept with him in the cockpit. He set a new speed record between the desolate landing strip known as Bidon V and the Algerian town of Reggane. Over North Africa, frigid gales tossed the craft back and forth until he had to bite the fingers of his right hand to keep them from freezing. He ran into thick curtains of airborne sand above the Sahara, then billowing snowstorms. "The mountain peaks are covered so deep that roofs are all that is visible of houses," he wrote in his typically evocative prose. With the plane pointed north, the engine spewed oil at an alarming rate. Yet another sandstorm spun out a series of tornadoes, one of which tore at the plane's canvas wings before Cukurs managed to land near a military outpost. On the ground, the fierce winds threatened to pull the aircraft apart. As the *Duchess* bucked and twisted on its wheels, Cukurs jumped out of the cockpit, sealed up the exhaust pipe with rags, then threw his body on the rear fuselage to keep the plane from being thrashed to pieces. He was left bruised and exhausted, his mouth and goggles packed with sand.

Nine months after leaving Latvia, the aviator spotted the airfield in Liepāja, where thousands of people cheered and waved as he motored his plane down from the clear sky, guided to the runway by three roaring bonfires. When he put the *Duchess* down, the crowd, wild with joy, lifted him from the cockpit and carried him around the airfield on their shoulders. On national radio, the minister of war announced that Cukurs was being promoted to captain "for spreading the name of tiny Latvia far and wide." The government granted him an estate in the parish of Bukaiši; he was also awarded the Harmon Trophy, given annually to the world's outstanding aviator, and inducted into the Order of Santos Dumont, an honor bestowed only on the world's top pilots.

Herbert Cukurs' insatiable hunger for glory had made him a household name.

Cukurs was a dynamic speaker, and the men and women in the crowd that night in 1939 sat rapt as he described his adventures. But despite the drama of The Gambia journey, the highlight of the evening was still to come. Earlier that year, the aviator had made a journey to the Middle East. The screen behind him lit up as black-and-white photographs of desert scenes flickered to life.

The flight unfolded against dark clouds descending from the west. The news from Berlin was increasingly bleak. Hitler had been named chancellor six years before, and all political parties except the National Socialists had been banned. The Nuremberg Laws, including the Law for the Protection of German Blood and German Honor, which prevented marriage and sexual intercourse between Jews and non-Jews, had gone into effect four years before. Jews were already being detained and murdered by the Gestapo, though the systematic killing hadn't yet begun, and the Wehrmacht had marched across the Polish border only three months earlier. In such a fraught atmosphere, Palestine seemed an odd destination.

The aviator told the crowd about his preparations for the journey. He'd packed medicine for typhus, snakebite, and *E. coli*, and added dried pieces of rye bread "so as to remember the fields of my dear homeland." He stuffed a big Mauser rifle in the cockpit to fight off bandits and a Leica camera to take the pictures that now clicked into place behind him. His 2,900-mile expedition, which he planned to complete in segments, would be the first such journey undertaken in a self-built aircraft. The trip was as dangerous as anything he'd ever attempted; the weather as he set out was miserable. "I've never encountered such a chaos of incessant thunder rolls, the wild melody of rain howls, and the wails and whistles of wind," he wrote. An airliner nearly ran him down over Berlin, and, later on, when he looked down into the bottom of his cockpit, he saw black oil bubbling across the aircraft's floor. The propeller spun unexpectedly one morning as he was trying to start the engine and sliced his right hand down to the bone.

After many near misses and mishaps, he made it to Palestine, which he described to the audience in rapturous detail. Yoel Weinberg, a studious young Jewish man who would go on to become a biblical historian, listened to the Latvian describe the country and its people. "I remember Cukurs speaking with wonderment, amazement, even enthusiasm, of the Zionist enterprise in Israel," he said. Audience members leaned forward in their seats to study the photos. Here was Tel Aviv! And the birthplace of Jesus! But Cukurs had also wandered away from the tourist trail, to the citrus groves of Petah Tikva, the "Mother of the Moshavot," established in the late 1800s with a grant from Baron Edmond de Rothschild. And to Rishon LeZion, on the coastal plain, which had been mostly sand and scorpion holes when ten Ukrainian Jews arrived in 1882, bought the land from the local Arabs, dug a well, and created a coat of arms. (Their motto, "We have found water," was a quote from Genesis.) Now the city boasted orange trees, a winery, and three thousand pioneers.

The aviator finished his speech; the screen behind him went black. Waves of applause poured down from the darkened balcony; the talk had electrified the audience.

Cukurs gave other speeches around Riga that year; it was one of the ways he made his living. Perhaps his main motivation that night was simply to squeeze a few lati (the Latvian currency) out of Riga's Jews, who were always dreaming of the Holy Land and always getting overexcited when some goyim talked about it in even vaguely positive terms. But his trip to Jerusalem and his very presence in the Jewish Club in the winter of 1939 hinted that Cukurs was a new kind of European, at least partly free of the old hatreds. The news from Berlin was certainly grim, but the men and women who filed out of the double doors after the talk could take some comfort in the fact that this national hero had taken the time to meet with them publicly. "I had been studying at Hebrew school at the time," said Yoel Weinberg, "and Cukurs' tales fired my imagination." When Weinberg woke up the next day, he raced to Bible school to talk about the aviator with his friends.

TWO

Zelma

IT'S DOUBTFUL THAT NINETEEN-YEAR-OLD Zelma Shepshelovich was at the Jewish Club that night. She was quite poor in 1939, a college student living in a closet-like room in Riga. To pay her tuition, she knit gloves for the ladies of rich families in the capital or occasionally took a job in a factory. Zelma's dream, which she felt she could tell no one — not her family and not her classmates — was to finish her studies and go to Palestine, where she could live among other Jews. But Zelma most likely heard about the lecture from her friends in the city, who often came to her room to talk for hours about their favorite things: art, serious books, the great world outside Riga.

Zelma had grown up in a prosperous German-speaking family in the small Latvian town of Kuldīga — prosperous, that is, until the Depression struck. She'd come to the English College in Riga a year before to learn foreign languages after passing the special examination that all Jewish students were required to take. She was bright, fiercely outspoken, and beautiful. Her rich auburn hair (or "lion's mane," as her friends called it) set off a face that resembled a Jewish Rita Hayworth's, but there was an imperiousness in her eyes and a slant to her brows that spoke of her fiery nature.

The Riga she found on arriving from the countryside was small, the capital of a country you could traverse by train from north to south in just a few hours, but it had its own charms. During the winter, farmers pushed their sleds, loaded with corn and potatoes to sell in the old city, up the frozen Daugava River; in the summer, lovers escaped to the islands that lay in the middle of that river to lie under the hot sun and kiss among the lindens. On the streets, peasants maneuvered their carts past women clad in dresses designed in French ateliers, and small boys pestered the train conductors at the main station for the stubs of long-distance tickets to Vienna and Istanbul, which they added to their collections. Some called this city of 350,000 souls "Little Paris"—but that's what the Hungarians called Budapest, too; most small European cities saw themselves as being a bit more cosmopolitan than they really were. Still, there was a deep culture here, much of it Jewish. The Riga Conservatory was a producer of musical talent on a par with many more famous institutions. As the great violinist Isaac Stern once said, if you met a Russian violinist, it didn't matter where they were born, they'd all come through Riga.

Zelma had long dreamt of what her university days would be like, sitting in one of the cafés on Artilērijas Street and having meaningful conversations about life and art with professors who, unlike her friends, had actually been out of the country. But her first months in Riga were rather lonely. Her fellow Latvian students were hard to talk to because, she sensed, they didn't like Jews. No one had to tell Zelma that there was anti-Semitism in Riga, as in every part of Latvia; she'd felt its bite more than once. "The best compliment you could hear was, 'Oh, you're not like a bloody Jewess, you're like one of ours,'" she said. One Jewish child, about the same age as Zelma, bore on his forehead a scar where an angry Latvian boy had hit him with a stone for being a "crooked beak." Another was denounced for supposedly cheating at a game: "I came home. I felt, I still have the feeling, a terrible feeling of being accused of something I was innocent of and being called a dirty Jew in the bargain." A Jewish schoolboy once turned to see a classmate making a pig's ear out of a piece of cloth; the other boy then pointed it at him and taunted him with it.

Jews were casually referred to as "the black growth." The city's lone Jewish policeman was such an unusual sight that people ran up to him on the sidewalk just to gawk. Were rabbis beaten on the street? Were the temples defiled? No, nothing like that. But if you were a Jew who wanted a job in a Latvian-owned factory, you knew better than to apply. Between the Jews and the Latvians, there was often a chill and a silence where one might have desired warmth.

Only a minority of Jews, however, believed that the prejudice went as deep as that found in Berlin or Warsaw. The mentions of Jews in Latvian folk songs were, for the most part, refreshingly neutral, even affectionate. Only a portion of the Russian *Protocols of the Elders of Zion,* the fabricated anti-Semitic document that purported to reveal Jewish plans for world domination, had been published in Riga. One historian wrote of the lands that had become Latvia, "Of all the territories of the western reaches of the Russian empire . . . those of Courland and Livonia were almost alone in their absence of pogroms." In fact, Jews persecuted in Germany, Poland, and Lithuania during the 1930s actually made their way to Latvia to find sanctuary; it was known as a refuge. In hindsight, it's easy to mock such distinctions, but they were all people had to go on at the time. "What they felt in their hearts," one young Jew said about the Latvians, "I don't know."

To be fair, the Latvians might have made a similar claim about their Jewish neighbors. Many Jewish families saw themselves as culturally German or culturally Russian, not Latvian. "Anything which came from Germany was 100 percent wonderful," one young woman recalled. Indeed, what did it even mean to be culturally Latvian? What great poets or composers had the country produced, what epics, what rivals to Pushkin or Rilke? Even with regard to language, one of the great markers of life and status in Riga, the native tongue was often disdained in Jewish homes as a kind of bastardized Russian, even though Latvian is ancient, older than any Slavic, Germanic, or Romance language. A good German Jewish child learned to say *danke* and *Fräulein* before she ever learned Latvian, and many never bothered to learn Latvian at all. One Jewish woman remembered being told as a girl that the language "is for speaking to the dog and the maid."

Jews were deeply invested in the independent Republic of Latvia and its hopes. They established businesses, became doctors and engineers, and founded the city's leading import-export houses; most strove to be good citizens. They loved their country and wished to be admitted to the institutions that were closed to them. But did they feel the same kind of ecstatic bond with their native *Kultur* as a Berlin Jew did to his? No, they didn't.

Zelma's experiences were fairly exceptional. She'd encountered ridicule from the Latvians, which was common enough, but she'd also known deep love. Her nanny, Ieva, had been with the Shepsheloviches since before Zelma was born and was treated like a favorite aunt. When Zelma's parents lost their two delicatessens during the Depression and became "penniless, absolutely penniless," Ieva refused to leave. She suffered along with the family. When Zelma was two and three years old, the nanny would take her out for walks, and the neighbors would kid Ieva that Zelma was really hers, not the Shepsheloviches'. They even went to church together, the Latvian spinster and the headstrong Jewish girl. "She loved me as her own child," Zelma recalled.

When Zelma began her university studies in 1938, Latvia was ruled by the authoritarian Kārlis Ulmanis, the former prime minister who'd taken power in a bloodless coup d'état four years earlier. Ulmanis outlawed all political parties, including the anti-Semitic Pērkonkrusts (Thunder Cross) organization, and assumed the role of a strict, benevolent father of the nation. One of his first edicts was to ban newspapers and broadcasts from the outside world, which had the effect of keeping many Latvians in the dark about the rise of Hitler. Under the slogan "Latvia's sun shines equally over everyone," Jews and other minorities were protected, even as Ulmanis pursued a "Latvia for Latvians" economic nationalism.

In the summer of 1940, however, the rather sleepy city that had become Zelma's home changed almost overnight. Latvia was absorbed into the Soviet Union under the secret protocols of the Molotov-Ribbentrop Pact (also known as the German-Russian Nonaggression Pact) between Stalin and Hitler, which had granted the three Baltic countries to the USSR in 1939. Ulmanis was forced to step down, and the nation came under the control of the Kremlin. The following June, Zelma watched as columns of impossibly young, red-cheeked soldiers marched into the capital by the thousands. The Soviet occupation had begun.

Late one evening, Zelma was walking back to her apartment when she saw white figures approaching her in the darkness, tall ghostly beings she couldn't identify. She stopped; the shapes drew closer. After a moment, she realized the creatures that seemed to float toward her were newly arrived Russian women, the wives of officers and other soldiers, who'd dressed in long gloves and billowing white nightgowns just looted from the local shops by their husbands. In their ignorance, the women had mistaken the lingerie for luxurious ball gowns and were now promenading through the city to show off their finery. A few of them sported two or three purloined watches on each wrist. Some Latvians who strolled by found the sight hilarious—these country Russians were so gauche! But Zelma was depressed by the spectacle. "It was a terrible sight," she said.

Far worse was to come. Latvian money immediately became worthless. Stores, farms, barbershops, and theaters were nationalized, their owners often given nothing in return. The new Soviet commissars confiscated private savings, dooming middle-class families to poverty; formerly prosperous businessmen were soon selling their household goods on the streets. Soviet troops, drunk on cheap vodka, broke into shops and pawed watches from their cases and liquor from the shelves. Enormous portraits of Politburo members, the new ruling authority in Latvia, were hung on the sides of buildings downtown. Books, including *Gone with the Wind,* which it seemed everyone in Latvia had been reading before the occupation, were banned, replaced by the works of Lenin and Stalin; entire libraries were emptied and their contents bundled up or pulped. Thousands of prominent people disappeared in the middle of the night and were driven deep into the USSR to starve in the forests or work in the labor camps of the Gulag. As the Bolsheviks ruthlessly eliminated anyone who might possibly challenge their rule, Latvia lost a full 2 percent of its population: priests, intellectuals, generals, moguls. The brightest lights of the country were put out.

Latvians were increasingly impoverished, traumatized, filled with inexpressible rage. "Just as intensely and deeply as we idealized Latvia," wrote one native author, "and we bestowed on its memory a virtual sanctity, we vilified its antithesis, everything Bolshevik. Anything Russian or communist represented the diametric opposite, the incarnation of evil, cruelty and infamy." As she walked to her classes, Zelma saw the paranoia and despair on the faces she passed. Her British professors plotted their escapes—"We are going to leave this goddamned country," one muttered to her—and even her friends wondered if they might be next on the list for Siberia. "People were scared," she said. "What is going to be forbidden in the future? Who is going to be exiled?" The Latvians began speaking of the Baigais Gads, the "Year of Horror."

One weekend, to escape the grimness of Riga, Zelma went home to Kuldīga. There she was invited to a "ball," just an open-air country dance really, not far from the Shepsheloviches'. Zelma didn't want to go, but to please her mother, she agreed to be escorted by a Jewish boy named Orele, the son of a neighbor. They walked along the country roads until they heard the sound of fiddle music and stamping feet, which led them to the party.

The two were dancing when a figure appeared among the couples. "All of a sudden, a tall gentleman wearing a Russian uniform, an officer's uniform, came up to us and asked Orele if he could dance with me," Zelma recalled. She was startled; Russian officers were demonized among the local population. "I didn't want it. But Orele said, 'That's alright.'" Zelma took the man's hand, and the two began to dance. She soon realized that her partner wasn't Russian after all, but a Latvian who'd just graduated from the military school.

"Do you live in Kuldīga?" he asked above the sound of the band.

"No, I live in Riga," Zelma said.

"Which street?"

Zelma thought this was quite forward. She gave him her address—6 Stabu Street—but not her apartment number or her family name. The dance ended. Soon after, the Latvian approached and asked for another. Zelma declined. "Can I take you home then?" the man asked. Zelma agreed, but only if Orele accompanied them.

It was an odd encounter. Except in a few cases, Jews and Latvians didn't mix romantically; even Zelma, the rare Jew with Latvian friends, the girl who'd once been mistaken for a Latvian child, expected to spend her life with a good Jewish boy. She had a boyfriend, Max Schwartz, whom she had grown very fond of. "We were supposed to marry."

Zelma, Orele, and the Latvian stranger walked the country lanes to her home, and soon after, she took the train back to Riga, never expecting to see the man in the Soviet officer's uniform again.

THREE

First Night

IN THE SPRING OF 1941, news of Hitler's victories dominated the BBC News and Radio Moscow programs that many Latvians listened to in the evenings. In the pain and humiliation of the Soviet occupation, some of them began to see the Führer as a possible liberator; even a few Jews thought this way. But that spring, their faith was battered nightly, as reports of new anti-Semitic atrocities in places like Berlin and Düsseldorf came over the airwaves. Jewish refugees slipped into Riga with tales of their own. One young woman remembered an encounter with a German Jewish man. "He met my father and said, 'Do you have any money?' My father said, 'For what?' And he said, 'For you and your family to go to Israel, go to the United States, go anywhere you can, just run away from Latvia, because Hitler is going to come and he is going to kill you.'" Her father told the man to stop being ridiculous.

For those who didn't believe the stories, there was the *Brown Book*, an anti-Nazi tome written by the Soviet spy Otto Katz, originally published in Paris in 1933 and widely distributed after that. Refugees from Vienna and points west brought the book to Riga and passed it around, just as in other years they had brought American film magazines or opera bills. Inside were accounts of swastikas shaved into the hair of Jewish men, of books burned on German streets, and of Jews being hunted, beaten, and shot down "while trying to escape." Horrible things. Children in small Latvian towns flipped quickly through the pages to get to the graphic black-and-white pictures. "*That* put goose bumps on us," remembered one.

All of Riga talked about Hitler that spring and summer; one could hardly spend an hour at the Jewish Club without hearing a new theory about what would happen if his storm troopers came to the city. *Maybe he'll be strict with us,* one popular argument went. *Maybe we'll lose some civil rights, but nothing more than that.* Or: *Don't worry, Mr. Roosevelt is going to help us.* Zelma's father spent hours listening to the BBC and came up with his own idea as to what might happen if the Wehrmacht arrived. "In the worst case," he told Zelma, "they will make us work." He couldn't, in his heart, believe that Germany, the birthplace of Beethoven and Schiller, could actually be doing the things it was accused of. "This was the thought of all German-speaking Jews."

That summer, Abram Shapiro was a music-mad teenager. He was sixteen, looked twelve, and played the violin like he was thirty-five. His father worked in public relations for a large textile company and provided a comfortable life for Abram and his sister, Selma. In the winter, they went ice-skating and skiing, and in the summer the whole family decamped to the beach, where they would play volleyball on the golden sand. Their Latvian neighbors in the building at 4-4 Zaubes Street were friendly; Abram played cowboys and Indians with their sons. His relatives, all of whom called him "Pimpelchen" ("pimple" in Yiddish), were always at the Shapiro apartment, filling the rooms with music: Abram on violin, his mother on piano, his uncles singing with their lusty voices. "It was just a beautiful life," he said.

Abram knew that awful things were happening in Germany, but he wasn't especially concerned. Hitler was a noxious insect buzzing in the distance; people had to get on with their lives. The Shapiros weren't immune to the war; on the contrary, Abram's parents were housing refugees in their apartment, as many Jews did. The exiles did tend to prophesy darkly about the future, but Abram didn't pay them much attention. "I never imagined the war and Nazis would come to Latvia," he thought. "It's going to happen over there, it's not going to happen to us." Once in a while, the air-raid siren blared, and he and the other apartment house dwellers practiced running to the shelter in the basement. There his family gossiped and, if they had brought a violin, played music until an hour or two had passed, when everyone went back to their flats. Some children he spoke to in the basement were actually excited that the Germans might be coming. "I will no longer have to go to school" was the first thought of one Jewish student.

The memories of Riga's Jews carry an echo of the stories English people told of the weeks and months before World War I—the garden parties, the balls, and poet Philip Larkin's "never such innocence again." Many years later, one Riga Jew still found it difficult to understand his own behavior. "I am amazed by the incredibly carefree way we lived our lives," he said. "We remained unmoved . . . We believed we were far from the scene of action." The speaker was a boy at the time; certainly, many older Jews spent their nights anguishing over the question of what would happen to them and their loved ones if Hitler's troops stormed the city. Yet, in all the discussions, in the rumors and counter-rumors, there was hardly any mention of a crucial point: *If disaster does come, what will "our" Latvians do?*

When Hitler launched his attack on the USSR on June 22, 1941, each Jewish family in Latvia was forced to make a decision: stay or go. Some rushed to the central train station to board a train for Moscow or phoned the contacts they'd saved for such an emergency. Others encouraged their neighbors to stay calm. In the town of Jelgava, the director of the Jewish bank, a man named Jakobson, stood in the middle of the marketplace, his arms raised in the air, as Jewish families in heavily loaded wagons rushed through the streets, their horses' noses pointed eastward. He called out in a loud voice for his neighbors to come to their senses and return home. Many did. In Liepāja, a prominent doctor named Arkadi Schwab, who'd refused to believe the stories of German atrocities, was asked by his Latvian colleagues to join them in a welcoming committee to greet the Wehrmacht as soon as they arrived. He agreed.

In Riga, the Russians sent a car for the country's leading epidemiologist, hoping to spirit him out of the country before the Germans marched in. The Jewish doctor felt it was morally wrong to leave his patients and told the driver he was staying. When one Jewish family arrived at the Riga train station, scrums of shouting people pushed toward the carriages. The family had left the wife's mother behind, as her arthritis meant she could barely walk. As the passengers surged into the train, hauling suitcases and parcels of food for the journey, the wife turned to her husband and said simply, "My mother." He looked at her. "Well, let us go home. But you know what can await: concentration camps and raping and everything." They went back to their apartment to wait for the Germans.

The Jewish director of an industrial plant rushed to work one morning soon after the invasion to find that he'd been let go. He knew in that moment that he must leave the country. But when he hurried out of the plant to get his car for the journey east, the young Latvian chauffeur informed him that the car engine was "not in order." The director blanched. "This means they have sabotaged it." He began running home through streets where Russian soldiers nervously crouched behind barricades.

Across Riga, phones rang and rang in Jewish homes. Often it was a loved one or a colleague on the line, warning that the last trains for the Soviet Union were leaving within minutes. Were they coming or not? One young woman hurried through her family's apartment to the sound of the phone, gathering things as her family tried to coax her mother out of bed. She was feeling ill, but time was short. "I became so terribly distraught, not knowing . . . how to proceed, when suddenly our telephone went silent." The line had been cut.

Zelma was alone in Riga; her family—her ailing father; her

mother; her younger sister, Paula; and her brother, Zelig—were back in Kuldīga. They'd made plans to escape to the USSR; Zelig had even arranged for a car to evacuate them. But at the last moment, Zelma's parents decided they couldn't leave their daughter on her own, so they hurried to the city to fetch her. Once the Shepsheloviches were reunited, they rushed to the railroad station, hoping to catch one of the last trains out. When they tried to get on an eastbound carriage, they found that all the seats and all the compartments had been commandeered by Soviet officers; even the corridors were packed full of anxious passengers. "There was no way out," Zelma said. "We were stuck." Only Zelig managed to get away.

Abram Shapiro's family secured him and his parents a place on a Moscow-bound truck. But if they fled to the Soviet Union, tiny, music-loving Abram would be old enough to be drafted into the Red Army. A family friend warned against the trip: "You'd better go home. They're going to kill you out there on the battlefield." It seemed obvious that one choice—*stay or go*—would turn out to be indisputably better than the other. But no one could see into the future. The Shapiros decided that staying offered Abram a better chance of surviving; they let the truck leave and hunkered down in their apartment on Zaubes Street.

The last trains, packed to their roofs with luggage, prepared to leave for Voronezh and Moscow. The drivers of buses and trucks called out to their dawdling passengers, then revved their engines and rolled off in clouds of blue diesel smoke. Those Latvian Jews who remained stayed in their houses, playing cards to calm their nerves or slowly turning the radio dial looking for the latest news reports. A few pulled out books like *Gone with the Wind*—they'd been hidden from the Soviets in attics and behind walls—and reacquainted themselves with Rhett and Scarlett. One of the few blessings about the approaching German invasion was that one could, for a few hours at least, read what one wanted to read.

On the night of June 30, Zelma sat in the basement of her apartment building with her family listening to German bombs fall on the old city, astonished that the rumors they had mulled over for so many months had suddenly taken the form of something as real as the crump of high explosives. Perhaps for Zelma there were a few pangs of guilt, too. She was the reason her family was trapped in the city; the fate they would share had originally belonged only to her.

Hour by hour, the sound of artillery grew steadily louder as the Germans advanced. The next morning at 6 a.m., the guns fell silent all at once and Latvians watching from their windows saw Wehrmacht soldiers advancing carefully into the burning, rubble-strewn city. The corpses of Russian soldiers lay sprawled on doorsteps downtown. Latvian sharpshooters perched high in the nearby buildings picked off Soviet troops running fast through the eastern districts. One Jewish boy who'd gone out on the streets to see the Nazis arrive watched as the same Soviet unit appeared on his street again and again. The soldiers would charge down the block, turn the corner, then reappear at the other end minutes later and dash by again. The men had panicked and, confused by the city's layout, were running in circles.

After many hours of silence, familiar music now played on the radio: the national anthem, "God Bless Latvia!" People slowly emerged from their basements. It was a splendid morning. "The golden sun shone down from a cloudless sky," wrote one resident of the city. Jews came out of their homes to see the Nazis, those exotic creatures about whom they had heard so much. One boy ran from his apartment to Tērbatas Street, a main thoroughfare, and stood on the curb to await them. Almost as soon as he arrived, a German motorcycle regiment complete with sidecars came flying around the corner. The men's faces were covered in a thick layer of dust as they roared past in pursuit of the Russians. So these were the Nazis; he thought they looked stern but not much like monsters. Others came across "strong, tanned fellows with rolled-up sleeves," hugging the walls as they advanced neighborhood by neighborhood. Many already had flowers in their rifle barrels, put there by women celebrating their freedom.

Pent-up joy burst out among the Latvians; thousands of them streamed out of their homes attired in their traditional folk outfits of wool dresses and fine fringed capes, a sartorial signal that they believed the Germans had come to restore their independence. They tossed bouquets of daisies onto the mud-spattered tanks and embraced the young Wehrmacht soldiers. When one Jewish boy spotted German planes swooping low over his house, he waved to the pilots; they saluted back. "It was a strange excitement," another resident remembered. The portraits of the Politburo members were torn down from the buildings in the old city; red-and-white Latvian flags, which had been kept hidden during the Soviet occupation, now fluttered from almost every window. Latvia, it seemed, was now safe from the Bolsheviks.

The Jews who ventured outside that morning sensed a change in the air. They began to feel somehow exposed, as if the atmosphere had become slightly toxic to them. "I am walking very fast," the plant director whose car had been sabotaged wrote, "and realize that I must get home as soon as possible. Some of the Latvians eye me suspiciously. Some have a sinister, questioning look about them when they see me, as if to ask, 'Is he a Jew or isn't he?'" Latvians stopped pedestrians on the street and forced them to say the word *kukurūza* (corn) to see if they pronounced it with a slight rasp, a sign of a Yiddish speaker. Neighbors who'd greeted them the day before now walked past without a word. That morning, a Riga newspaper published a list of people exiled by the Soviets, bracketed by a thick black border. The names of hundreds of men and women were recorded there in columns that covered entire pages. But when Jewish residents went slowly through the list looking for their departed loved ones, they found nothing. What could it mean? Why had their daughters and uncles been left out?

When the radio stations switched away from the national anthem, the Jews got their answer. Now that the Nazis controlled the press, they focused relentlessly on disseminating a single false message: The Jews had been the handmaidens of the Soviet occupation. They'd welcomed Stalin's troops, who had raped and humiliated the country. They were the "inner enemy" and must be punished for their betrayal. Fleets of German bicyclists pedaled away from Riga along country roads, carrying in their satchels "proof" that the Jews had been Stalin's accomplices. Stories about Soviet interrogations ran in the dailies. "I awoke on the floor in another room with my hands and feet tied," read one account. "Next to me sat a red-headed, curly-haired Jew with a cigarette between her teeth. She looked at me coldly, and her hands held a long needle with a blood-stained tip. The red-haired scoundrel asked me: 'Will you confess or not?' . . . She slowly pushed her horrible needle into my thigh. Powerless, I yelled for help, but a Jew-Chekist with a wet stinking rag gagged me."

The Latvian-language newspapers reprinted speeches by Hitler and the Nazi propagandist Joseph Goebbels. One paper, the *Nacionālā Zemgale*, published in the city of Jelgava, reached heights of anti-Semitic viciousness in its editorials "which perhaps has no parallels in Nazi propaganda anywhere, not even Streicher's *Der Stürmer*." Soon after the Germans arrived, the editors wrote: "In the world there is nothing lower, nothing more dastardly, than the ass's kick to the wounded lion, and that is what the Jews did to the Latvians . . . We are now secure behind the shield of Adolf Hitler and the great German Army. The murderers and looters must receive their due punishment . . . No pity and no compromise must be shown. No Jewish tribe of adders must be allowed to rise again."

Latvians claimed they'd seen Jews kissing Soviet tanks when Stalin's troops had marched in a year earlier; this rumor quickly attained the status of official history. For their betrayal, one prominent journalist wrote, the Jews "should die as a nation."

There were small signs that caused some Jews to hope that the storm of hatred and accusations would quickly blow over. At the appointed time, Arkadi Schwab, the Jewish doctor who'd agreed to welcome the Germans with his colleagues, marched out to the town limits and shook hands with the German officers. Schwab asked about the treatment of people like him under the new regime. A lieutenant assured him that the Jews "had nothing to worry about." Schwab returned to his family and told them the good news. But most Jews felt suddenly isolated, mysteriously threatened. "A sea of hatred surrounded us," one man recalled.

Jews waiting in their apartments that morning heard the sound of boots in the stairwells. In the intimate geography of the hallways, a listener with his ear to his door could tell which floor a visitor had stopped at, even which apartment. That day, each time the boots stopped, it was at a flat belonging to a family of "crooked beaks." Then came several knocks, louder than usual; the Latvians were using the butts of their rifles to pound on the doors, sending a hard, sonorous report through the building. After the pounding had stopped, a voice would call out: "Are there any Jews here?"

Zelma, sitting with her father, mother, and younger sister, along with the Latvian nanny, Ieva, heard the sound of footsteps stop outside their apartment. Someone banged on the door.

Ieva went to see who it was.

"What do you want?" she asked the soldiers standing outside.

"We have come to pick up the Jews."

"There are no Jews living in *this* apartment," Ieva huffed, and closed the door. After a moment or two, the sound of boots moved off down the hallway.

Zelma listened. The footsteps reached the stairwell and moved up, emerged one floor above, and went to the apartment above the Shepsheloviches', where a Jewish family, the Kottens, lived. The sounds of a rifle on the door. Then voices.

"Kotten," a soldier's voice said, somewhat perplexed. "Kotten is a Latvian name or a Jewish name?"

When those Jews who answered the knocks on their doors were pulled from their homes, they saw Latvian volunteers, dressed in a hodgepodge of uniforms but all wearing the same red-and-white armbands. It had been less than twenty-four hours since the Nazis had arrived, and yet the Latvians had their markings ready. Some of the soldiers carried rifles, others handguns. You could smell the vodka on them; many were dead drunk. "On the first night," remembered a Jewish woman who was a teenager at the time, "there was already a knock on our door. I opened it to a small gang of Latvian youths, 16 to 17 years of age, gathered on the staircase. They were led by a neighbor of ours, whom I knew well. Since my arrival in Riga . . . he was always doffing his hat in flattering greeting at a distance." But now the man said nothing, only pointed out the girl's father. The visitors took him away.

Jews were stripped naked and forced to march through the streets wearing military caps taken from dead Soviet troops and singing Russian ballads. Some were told to carry ten-ruble bills, which featured a portrait of Lenin, between their teeth. The Latvians stopped occasionally to beat their prisoners with anything that came to hand — police truncheons, logs — until the men's faces were swollen and covered in blood. The guards shouted, "Jews! Bolsheviks!" and "Stalin's unconquered army is in retreat!" In the small town of Preili, Jews were forced to wear clown masks and sing, as if they were on their way to the circus.

Yiddish-speaking prisoners were taken to fields and backyards and ordered to dig up the bodies of Latvians shot by the Soviets, using their bare hands; many of them contracted tetanus from the decaying flesh. Some of the corpses were found with their genitals cut off or missing eyes. The Jews, innocent of the killings and having lost thousands of their own in the purges, were photographed alongside the bodies as if they were unrepentant murderers being reunited with their victims. The pictures ran in newspapers across Latvia for the next several weeks.

In their apartment at 4-4 Zaubes Street, Abram Shapiro's father refused to believe the stories of terror that were just now beginning to spread through the city. He announced to his wife that it was time for him to return to work. "My mother was like, 'Don't do this,'" Abram remembered. But Pinchas Shapiro was a war veteran who'd served honorably as a freedom fighter in the Tenth Aizpute Infantry Regiment. He was as Latvian as the next man. He took young Abram with him to the factory.

Together they walked through Riga's streets; when they arrived at the plant, they found it guarded by a group of Latvians, who eyed them with suspicion. As the workers lined up to enter the factory, the new guards shoved the Jews out of the line. A bus pulled up, and Abram and his father were forced onto it. The bus took the road to the outskirts of the city where some of the fiercest battles between the Germans and Soviets had been fought. Abram saw the mud-covered limbs of dead soldiers jutting up from the ground like stunted trees. He and his father were ordered to carry the corpses to the truck and hose them off. The bodies were beginning to rot; the smell was overpowering. The Jews worked all day until the late sunset. "We were completely exhausted," Abram remembered. By now, Abram sensed the truth: "If you are a Jew, you are doomed."

The next night, Abram was sitting at home with his father, who'd decided not to go to the factory that morning, when they heard a knock. Mr. Shapiro opened the door, and a group of Latvians burst into the apartment. At their head was a craggily handsome officer wearing a heavy handgun strapped to his waist in a wooden holster. His muscled chest smoothed the wool of his uniform; his calves strained the leather of his boots. Abram's father knew him well. It was Herbert Cukurs.

Cukurs told Pinchas Shapiro that his family would be moving out of their apartment and would have to squeeze in with their neighbors on the floor below. "He liked our flat," Abram remembered, "and he wanted to set up his living quarters there." In addition, Pinchas would have to leave with Cukurs. Abram's father began remonstrating with Cukurs; he protested that he was a patriot who'd served his country on the battlefield. As Pinchas pleaded with the intruder, Abram glanced out the window and saw his Jewish neighbors forming lines down in the courtyard below; some were being clubbed to the ground, beaten with a violence that took his breath away. Abram turned back. "Oh, no, I'm a Latvian," Abram heard his father say. The elder Shapiro hurried to get the certificate showing that he'd fought with the Tenth Aizpute; he emerged from the back room holding the piece of paper in front of him and presented it to the Latvian aviator.

Cukurs glanced at the certificate.

"You're a Jew," he said. "Get out there."

The Shapiros were terrified. "We begged him to leave Father at the apartment," Abram said. But Cukurs turned away without answering. Pinchas put on his overcoat; he went up to Selma and Abram and kissed them. Abram stood next to his mother and watched as the Latvians led Pinchas out of the apartment into the hallway. Abram, protected by his youthful looks, felt the sudden desire to go with his father.

Hundreds of Jewish men were rounded up that evening. No one knew that night where they were going, only that they were being herded into covered trucks, which drove off without any announced destination. Zelma's father was spared, but her uncle and cousin were taken away. "They never saw daylight again," she said. It wasn't actually true. Most Jews survived that first night. But twenty miles away from central Riga, to the east, in the sprawling Bikernieki forest, thick with Norway spruce and looking-glass lakes, Latvian peasants spotted Soviet prisoners of war marching into the woods with picks and shovels in their hands. Over the next few days, large pits appeared, hidden from the roads by the branches of tall, verdant pines.

FOUR

Gogol Street

THE GERMANS, NOW IN CONTROL of the country, declared martial law. The Latvians were ordered to turn in their weapons; all native political organizations except for the radical right Pērkonkrusts were banned. A curfew was introduced. Special Jewish edicts began to be issued almost daily. July 3: *All Riga Jews must be removed from employment.* July 5: *Jews in Zemgale must leave their places of residence.* July 5: *Jews are forbidden to talk to non-Jews, on penalty of death.* July 9: *Prisoners of war and Jews who've remained hidden must turn themselves in.* Jews were required to walk in the gutter and not on the sidewalk, were allowed to shop only from 10 a.m. until noon, and were banned from attending public gatherings or walking on the seashore. They were ordered to turn in all radios; bicycles, cars, or other means of transportation; uniforms; and typewriters. Jewish women weren't allowed to wear hats or use umbrellas. Some Latvian collaborators devised unique punishments all their own. In the town of Bauska, Jews were taken from their homes and castrated.

The orders for the new laws and the violent subjugation of the Jews came from Berlin; the Latvians were worker bees, henchmen — never decision-makers. In 1941, Heinrich Himmler, the *Reichsführer* of the SS (the Schutzstaffel, or paramilitary arm of the Nazi Party), and

Reinhard Heydrich, the chief of the Reich Security Main Office, were slowly, in fits and starts, evolving a strategy for making eastern Europe *Judenfrei:* free of Jews. The German officers in the field — in Latvia, it was the brutal SS commander Friedrich Jeckeln — experimented with different methods, trying one thing and then another, searching for the right balance of lethality and efficiency. Was starvation better than a bullet in the back of the head? Was exile preferable to working the Jews to death? (Hitler had once imagined Madagascar might be a final destination for Europe's Jews.) What about portable gas vans — would they be worth the investment in equipment and fuel? Nobody knew the answers to those questions.

Himmler and Heydrich did, in 1941, have a pet theory: they believed the Latvians and the other eastern Europeans would do most of the dirty work for them. Once German forces entered the eastern territories and freed the native populations from their Jewish overlords, they would rise up and kill their oppressors in a bloody, spontaneous "self-purge." "No obstacles are to be placed in the path of the self-cleansing desires of the anti-communist and anti-Jewish circles in the newly occupied areas," Heydrich wrote on June 29 from Berlin. The Germans were convinced that anti-Semitism lay in the heart of every right-thinking Latvian. Himmler felt that by encouraging the Latvians to follow their natural inclinations and inciting them with propaganda that equated Jews with Bolsheviks, he could unleash thousands of killers on Jewish homes. This would pay a double dividend. The world would see that it was the Latvians and not the Germans who were committing the atrocities, sparing the Reich any more blotches on its international reputation and saving their own soldiers from unnecessary trauma. Himmler frequently urged his commanders "to take personal care that none of our men who had to fulfill this difficult duty ever become brutal or suffer damage to their psyche or character." If it was men like Herbert Cukurs firing the bullets, so much the better.

Jeckeln had made his name in the Ukraine, especially at Babi Yar, where he had organized the two-day massacre of 33,000 Jews. Transferred to Riga, he was competing with his fellow commanders for Himmler's approval. In turn, Himmler was competing with others in Hitler's inner circle, especially Hermann Göring, for access and power. He wanted his organization, the SS, to rule in the eastern territories and arrived on "absolute ruthlessness" as his guiding principle. The campaign in Latvia, which had no history of pogroms, would be a showcase for his particular style of extermination.

Soon after the occupation began, a Riga man was walking home from work when he saw a group of captured Russian soldiers being led down the street by a young German wearing a swastika armband. If anyone had it as bad as the Jews, it was the Russians. "They were barefoot, ragged and starving," the man remembered. Some Latvians, along with a poorly dressed Jewish woman holding her child by one hand and carrying a loaf of white bread in the other, stopped to watch the procession. Moved by what she saw, the woman gave the loaf to the child and gestured that he should hand it to one of the prisoners. A German guard saw the child approach a prisoner, pulled his rifle up, and shot the woman, her child, and the Russian soldier in turn. The Latvians cheered.

Jews stayed indoors as much as possible, listening for unfamiliar voices outside their doors. "From every corner of the building could be heard the shrieks and moans of Jews being beaten by the sadistic Latvians," wrote one woman. "People were rolling in their own blood." Jews invented doorbell codes. *Ring twice, wait, then once more,* they told their friends and loved ones. If the doorbell rang once, they didn't answer. "The whole thing was so surrealistic," recalled one Jewish woman. "It was like a play." On the streets, some Latvians spat at passing Jews being herded away from their homes. "People became worse than animals, blood-thirsty, without any pity," remembered a Riga resident about his Latvian neighbors. "And I must get the record straight — the overwhelming majority of them just couldn't care less."

On July 1, an all-volunteer force of Latvians was organized under the command of a handsome thirty-one-year-old ex-lawyer and anti-Semite named Viktor Arājs. The unit, which soon numbered about three hundred men, was authorized by the German *Brigadeführer* Walter Stahlecker, head of Einsatzgruppe A, a paramilitary killing unit that followed close behind the advancing Wehrmacht, identifying, hunting, and murdering Jews. The newly formed Arājs Commando began assisting in that work, its members always subordinate to the Germans. Herbert Cukurs joined the Latvian unit in July as a captain and began working as Arājs' second-in-command. Serving with Arājs offered valuable perks: it exempted the commandoes from being sent east to fight the Soviets, gave them access to valuables confiscated from Jews, and allowed them to sleep in their own beds instead of on a cot in a cold barracks. Cukurs was seen on the streets after the invasion, often drunk, almost always wearing a black leather coat, with his Russian-made Nagant pistol strapped to his waist.

On July 4, fifteen or so commandoes gathered in front of the Great Choral Synagogue on Gogol Street; some eyewitnesses would later claim that Cukurs was among them, though he would vigorously deny it. Inside were an unknown number of Jews who'd sought sanctuary in the temple; some said they were Lithuanians, others Latvians who'd failed to make it home before the curfew. Viktor Arājs arrived outside the temple and addressed his men. "Since the people of Riga hate the Jews," he shouted, "we must demonstrate our position by setting fire to the synagogue, so that nothing of the Jewish culture remains." The men went to their cars and siphoned gas into empty bottles, then made their way into the synagogue, where they grabbed the holy books and tore the pages out, throwing them on the floor. Holy scrolls followed, ripped and trampled underfoot. The men took their truncheons and smashed the glass in the windows and knocked the religious artifacts to the floor. Then they piled the debris together and poured gasoline over the pile. The synagogue was built of stone and wouldn't burn without fuel.

One of the men lit the pile, and the flames rippled in the dark interior as the Latvians withdrew from the building. Yells and shouts erupted from inside. "Devout Jews from the vicinity ran to the prayer house and tried to do something to fight the flames," wrote one Riga Jew. "They were insulted, humiliated and beaten back." Smoke began to pour out through the broken windows; the faces of the trapped appeared there and at the door. The commandoes shot at them with their pistols. Firefighters arrived, but Arājs' men told them only to spray down the nearby buildings so they didn't catch fire. The Jews inside cried out to the watchers, begging to be saved. "The bloodcurdling screams . . . pierced the flames as they burned inside the building." The roof collapsed with a clatter of stone and tile, and the fire grew hotter. How many Jews were burned alive or succumbed to the acrid smoke is impossible to tell.

"Life is indescribable," one Jewish man reported. "Terrifying days and nights have begun. We hear a door open or footsteps in the hallway; we hear them questioning the custodian . . . 'Which ones are the Jews? Which ones are not?'" Jews developed stutters or fell into suicidal depressions. Some Jewish girls began dressing in old clothes; they stopped combing their hair, as the Latvians snatched the most attractive women to rape and murder. Arkadi Schwab, the doctor who'd been assured that the Jews would be well treated when he went out to welcome the Wehrmacht, was arrested and viciously tortured; during their interrogation, the Germans gouged out one of his eyes. When Schwab stumbled across his prison cell trying to retrieve it, he was shot dead.

Greed quickened the Latvians' appetite for violence. Soldiers chopped off the fingers of Jews to take their rings, and there were rumors that civilians who'd been entrusted with valuables by their Jewish neighbors would offer them poisoned food and encourage them to partake. Max Kaufmann refused to let his son eat the provisions their neighbors arrived with, even though the two of them were "half-starving." Later, stores selling Jewish furniture and clothing sprang up on Riga's boulevards and side streets. Kaufmann passed one: "As we went to work in the gray light of early morning, we saw long lines of Latvian women who had come to shop." Some of the items were dashed with their neighbors' blood, but the Latvians stood in line for hours to get their chance at a bargain. Those Jews who ventured out noticed a large number of trucks on the roads, many of them filled with household goods: beds, chests of drawers, oil paintings, dining room tables. It was as if half the city had suddenly decided to move. But there were no families inside the cabs of the trucks, only soldiers.

For Cukurs and his fellow commandoes, those weeks in July were a kind of magical boy's adventure, where vodka was available for the asking, cigarettes were free, the best-looking Jewish girls were theirs for the taking, and there were parties almost every night. If they needed money in addition to their salary, they could simply sniff out the lair of a rich Jew and cart away his possessions. Jews were everywhere. They could be snatched from their homes or off the streets or tramcars; the conductors occasionally spotted one and called for the fugitive to be arrested. They tended to gather in certain spots. If a commando was hungover or tired and didn't want to go looking for "heads," he could simply pull up to one of the few grocery stores still allowed to serve them and "harvest some Jews" by ordering them into the truck. Riga was wide-open. Weeks before, many of the commandoes had been unemployed and at loose ends. Now they went about the city like medieval lords.

In the days after Cukurs arrested Pinchas Shapiro, Abram would go out on the street early in the day carrying a bag filled with sandwiches. Trucks with Jewish men standing in the back roared up and down Riga's boulevards. "Their faces were coal black with dirt," he remembered. Abram would toss sandwiches to the men as the vehicles chugged past, hoping against hope that one would somehow reach his father. If not, another Jew would be fed.

Cukurs soon returned to take possession of the Shapiros' apartment. The family had gathered their belongings in boxes; the furniture and their beloved piano would stay. Abram's mother asked Cukurs about her husband. Where was he? Was he safe? The Latvian pulled out a list from the lapel pocket of his leather jacket — these lists were carried by some Latvian and German officers — and ran a finger down it. Yes, here he was, Pinchas Shapiro. Cukurs informed Mrs. Shapiro that her husband was dead. Listening to his voice, Abram felt as if he were ninety years old instead of sixteen, so old and frail that he might not live past tomorrow. "I had made up my mind that every day was a gift from God," he said.

The Shapiros crowded into one of their neighbors' apartments on a lower floor. One night, Abram was sitting with his mother in their new flat when someone knocked on the door. It was a member of the Arājs Commando; he told the teenager that Cukurs wanted to see him. Abram followed him up the stairs to the family's old apartment.

As he entered his former home, he could see that there was a party under way. The sound of male voices spilled out of the interior rooms, and men in uniform were splayed on the Shapiros' couches, drinking vodka out of small glasses, their faces relaxed and smiling. Other commandoes stood around telling stories in Latvian as their companions laughed and glanced at one another in amusement. Among the scrum of uniformed bodies, Abram spotted Cukurs, who motioned the teenager over. When he reached the captain, Cukurs said that he'd heard that Abram was a talented musician. Cukurs pointed to the piano in the living room, the Shapiros' grand, and ordered the teenager to perform for his friends and fellow commandoes.

Abram sat down on the piano stool and placed his fingers on the keys. He began to play. The party continued, with the notes of the piano interlaced with the loud voices and the braying laughter. Every so often, Abram would look up from the keyboard and see Latvian soldiers and policemen on the sofa and chairs, the same sofa where his uncles — most of them now dead, he suspected — once sat on birthdays and Jewish holidays. It seemed impossible that this could be happening in the room whose every corner held for him a beloved memory, that his relatives were gone and had been replaced by these red-faced men.

After a while, Abram sensed a new presence in the apartment. He looked up and watched as a dark-haired Jewish girl—he didn't recognize her—was led out of the kitchen. The Latvian irregulars, led by Cukurs, gathered around the young woman. They began to clutch at her dress, feeling the flesh beneath it. "All the Latvians began having fun with her," Abram remembered. In a few minutes, they began to undress the girl. Someone pushed her down. As Abram continued to play his Brahms and Mozart, he caught glimpses of naked bodies through the scrum. "I saw with my own eyes how those Latvians raped her one by one," he recalled. The men violated the girl to the shouts of their comrades and Abram's piano playing, which he dared not stop.

On July 21, Riga's Jews were informed that they would be moved into a ghetto in Maskava Forštate (the Moscow suburb), a poor, sixteen-block neighborhood where before the war a middle-class Riga resident "would have been afraid to pass by." Furious planning began. Jews stood in line to find out where they would be living. "Completely impoverished and barefoot women," wrote one witness, "stood next to elegant Jewesses who were still wearing clothes that had been made in Paris salons." Movers robbed the Jews on their way to the ghetto, and at the entrance Latvian guards stole from them again, hitting them in the face with the stocks of their rifles if they protested.

The Germans had shuttered the English College; Zelma was now working to buy food for her family, as her parents were too ill to earn anything. One afternoon when she returned from work, her father was waiting at the door. "A tall Latvian has come and he was asking for you," he told her. The man seemed important—he was carrying a portfolio of some kind in one hand—but seemed confused about who the Shepsheloviches were.

"Do Jewish people live here?" he'd asked Zelma's father.

"Yes," her father said. "We are Jews."

"Is Zelma your daughter?"

"Yes."

"*Your* daughter?"

"Yes."

After her father finished the story, Zelma grew worried. What could this Latvian want? Had she ended up on a list? She couldn't think of who this man could be, but in this new climate, any unexpected visitor was worrying.

One evening soon after, there was a knock on the door. Zelma went to answer it. She guessed the man had returned. "I was prepared to die," she said, "[but] I wanted to spare my parents."

When she opened the door, she saw a familiar figure standing there: the young Latvian man who'd asked her to dance at the country ball in Kuldīga. "I was surprised," she said. "I hadn't given him the number of the apartment, and it was a huge building."

What the man said also caught her off balance. "I have come in order to see if I could help you," he said. "Is there anything you need?"

The risk the man was taking barely registered with Zelma; he could have been shot simply for asking a Jew such a question. Instead, another idea rushed into her head. This Latvian, who clearly found her attractive, was offering her charity—or perhaps he wanted to be repaid for his help in other ways. The thought repulsed her. "I was very angry and very proud."

"Thank you very much," Zelma told him. "I don't need anything."

The tall man didn't turn away. "The old city of Riga has been bombed and I have lost my furnished room," he said. "Could you perhaps help me find a room among your acquaintances?"

Zelma's expression softened. "I thought, *That's a miracle. All Latvian officers go into Jewish apartments, rob the apartment and kill the people. Here is a Latvian standing in front of me asking the poor Jewish girl to get a room for him.*"

She felt her animosity subside. "I think I may be able to help you," she said.

The man's name was Jānis Alexander Vabulis. He was twenty-seven and had gone to work for Riga's Buildings Department after graduating from the military academy. In short order, Zelma found him a room with a neighbor, Mrs. Joffe, whose husband had been murdered the first night of the roundups. The man moved in, and on Mrs. Joffe's door he wrote a message: "Defended by Jānis Alexander." No Latvian policeman or commando could enter the apartment to rob or assault her without the threat of consequences. It was a very unusual thing for a Latvian to do.

When Mrs. Joffe moved to the ghetto with her children, the man was alone in the apartment. He invited two young men he knew, Edgars Kraujinš, a school friend, and Laimons Lidums, to share the place with him. In the following weeks, Zelma's former nanny, Ieva, sent food from the countryside to the Latvian's apartment. Zelma would go out at night to pick up the food removing the yellow star (which all Jews were required to wear) from her coat to walk the short distance. She and her new neighbor would talk; he told her he went by the nickname "Nank," and that's what she called him.

In the ghetto, garbage removal and firefighting services were suspended, mail delivery was stopped, and telephone wires were cut. A separate sub-city emerged, one where possessing a newspaper filled with stories from outside was punishable by death. "A new world!" wrote the Riga resident Max Kaufmann. "A world full of cares and suffering." Latvian guards patrolled the perimeter. Jews couldn't approach the inner fence without endangering themselves; early on, two Jewish women who got too close were shot and fell onto the barbed wire. The guards also would fire at anyone who appeared in the windows overlooking the fence; the light of a single match could unleash a fusillade of bullets. The residents learned to keep to the interiors of their rooms.

Even non-Latvian Jews weren't exempt from the ghetto. When an American who happened to be in Riga for his sister's wedding got caught up in the evacuation order, he took his case to a Nazi police officer. The officer ripped up the man's passport and thrashed him with his baton until he bled. The man returned to the ghetto. A Czech Jew who failed to move behind the barbed wire on time went to the local police station to ask what his options were; he was taken out and executed.

In the Riga ghetto, musicians — some of whom had once performed at opera houses in Germany and farther abroad — played in the candlelit rooms. Their music would draw Jews from neighboring houses, who would sit, packed into the stairways, listening in silence. Births were forbidden in this new quarter, and Jewish doctors were forced to perform involuntary abortions. When one child, a boy named "Ben Ghetto," was born, he was immediately killed. Another infant was given a poisonous injection and soon died.

Riga became a pen where Jews were hunted for sport and profit, and Herbert Cukurs was an enthusiastic player in the game. One day, a young woman named Ella Medalje was staring out the window of her house when she saw the aviator driving through the gap in the barbed wire. He waved at the guards and pulled into the ghetto. "He was drunk and could hardly stand on his feet," she remembered. Cukurs opened the door, staggered out onto the cobblestones, and took out his pistol. Without bothering to arrest anyone, he aimed at the people walking by on the sidewalk, the barrel of the gun dipping up and down as he tried to steady himself. "Laughing devilishly, he started shooting the people like a hunter in a forest."

Only those with work permits were allowed to leave the ghetto. The work squads formed early in the morning and returned in the early evening, when the Jewish residents were checked by guards for smuggled food or contraband. One day, Reuven "Rudi" Barkan was walking in the ghetto after finishing a work assignment; he spotted Cukurs on the sidewalk ahead of him. The Latvian greeted a Jewish girl, ten or eleven years old, and asked in Yiddish if she would like a piece of candy. Cukurs, who'd grown up with Jewish neighbors and apprentices, spoke the language well. The girl said yes. "He told her to open her mouth. Then I saw him pulling his revolver, shooting [the] *maedchen* in the mouth and killing her on the spot."

At night, Cukurs and Arājs sometimes went prowling for Jewish women. They would select a house, march up the stairs, and rap on the door. Once, they "shouted that they needed coffee and girls." The older women would hide the younger ones under the beds, but they couldn't hold off the Latvians forever; there was always the chance that the pair would break in and start shooting wildly or drag everyone off to prison. Eventually, some girls would decide that the danger to their families was too great and would walk down to the street and get in the back of Cukurs' car, which was parked by the curb. "These girls never came back," said one ghetto resident. "It was common knowledge they were raped and shot." One woman recalled talking to a friend who was called to the scene of one of Arājs and Cukurs' orgies on Moskauer Street. Inside, the woman found the corpse of one Mrs. Schneider, who had been raped "in a perverse way."

Riga's Jews passed through a range of emotions: regret at not leaving, panic, depression, even frivolity. Some Jews spent all their money on fine food and liquor smuggled into the ghetto for outrageous prices. "People didn't want to take responsibility for the future," wrote one resident, "and lived only for the moment." Their moods were salted with the sharp taste of betrayal. Latvians and Jews had lived through the bitterness of 1940 together: the Soviet humiliation, the loss of their homes and loved ones. They'd suffered similar fates. And yet once the Germans arrived, their neighbors had turned on them with a ferocity that startled even the cynics among them. Where had it come from? "In my wildest dreams, I could never have imagined the hidden animosity the Latvians had for their Jewish neighbors," observed one Jewish man. "I had lived my entire life there among Latvians, who now considered me their mortal enemy and were prepared to kill me." Riga Jews hadn't experienced the slow, breath-stealing pressurization that German Jews had gone through beginning in the early 1930s. In Latvia, the terror had arrived all at once and, in many cases, at the hands of old friends. "The greatest tragedy was that these crimes were carried out not by strange, invading forces," wrote another Jew, "but by local Latvians, who knew the victims by their first names." For German-speaking families like Zelma's, who practically worshipped the land of Bach and Goethe, there was a double wound: the far betrayal of the Germans and the near betrayal of the Latvians.

That summer, a young man spotted a broad-shouldered man dressed in a green uniform walking in the ghetto. He studied the man's face. "I did not want to believe what I saw," he later said. "It was the well-known Captain Cukurs, who I had admired for many years." Cukurs, who could have stayed on the country estate awarded him upon his return from The Gambia and let the war sweep past his fence, became a symbol to many of the double cross that had snared the Jews. It was his face they remembered. "His presence lingered in my mind for a long time . . . ," the man said. "It was traumatic for me to realize that the flyer I had idolized had felt such vicious animosity toward the Jewish people." What dark urges, the Jews of Riga wondered, could have turned this hero into a man who loved to kill?

At that time, no one knew their relatives' final destination after the Nazis invaded their homes and dragged them away. There were only rumors of mass deaths; no graves had yet been discovered. But everyone in Riga did know that the first stop for the vanished was a certain house on Waldemars Street.

FIVE

19 Waldemars

EARLY IN THE NAZI OCCUPATION, the men of the Arājs Commando took over a large, luxurious home at 19 Waldemars Street, a three-story, buff-colored mansion that had formerly belonged to a Jewish banker named Schmulian. It became a killing ground with the air of a demented frat house. Late at night, the upper floors rang with laughter and deep-voiced toasts as shambolic parties raged almost until dawn. Lights blazed at all hours from rows of tall, oblong windows that overlooked the busy street. The Jews called it the Pērkonkrusts building, after the anti-Semitic political party.

Max Tukacier, who'd known Cukurs for thirteen years, was snatched off the street and brought with about two hundred other Jews, men and women, to the cellar of the building. The basement had a cement floor and walls streaked with blood; here and there were signs — a half-finished cigarette, a candy wrapper — that others had been kept in the cellar recently. On the door, someone had mounted a metal plaque that read CEMETERY OF THE JEWS. JEW STREET #2. In the cellar, some of the inmates stood, while others slumped against the walls and listened to the pounding of feet above, the sound of blows on bodies, the occasional bellow of pain. About every fifteen or twenty minutes, one of Cukurs' underlings would appear at the door, count off ten Jews, and take them upstairs.

After six groups had gone up, it was Tukacier's turn. He and the others climbed the stairs and were ushered into a large room with tall windows to the front. He saw discarded clothing—pants, dresses, suit coats—tossed haphazardly in a corner, along with many pairs of shoes. At a row of tables sat Cukurs and about twenty of his men. The Latvians ordered the Jews to kneel on the wooden floor in front of them. Once the Jews had obeyed, the men stood up and walked over. With a crack, one of the commandoes smashed a man to the floor with a truncheon; the others began attacking the kneeling Jews. "The Latvians, led by Cukurs, gave us a horrible beating," Tukacier remembered. Cukurs himself carried a rifle; he would walk around the room, spot an uninjured victim, and smash the butt of the gun into the person's face, snapping bones and spraying blood and teeth across the wooden floor. When he came to Tukacier, the Latvian raised the rifle in the air and brought it down on his nose, breaking it.

Cukurs called out for the Jews to undress and face the wall. The prisoners were shaking, some of them in shock, black hair hanging over cut brows and swollen lips. Tukacier staggered to the wall and took his place there. Cukurs walked behind the line, counting. He found eleven Jews, not ten, so he ordered Tukacier to the basement with four commandoes. On the way, the Latvians beat Tukacier "fiercely." In the midst of the blows and the grunting of the guards, Tukacier heard a number of shots ring out.

The young Jew was thrown into the cellar, where the others regarded him with fright. Thirty minutes later, Cukurs himself descended, crimson stains visible on the wool of his uniform. "Everybody upstairs!" he shouted. Twenty prisoners were brought to the large room, which they found empty except for the clothes of the ten Jews who'd last gone upstairs. The Latvians brought buckets of water and ordered the prisoners to clean the spattered walls and sop up the thickening pools of blood that had collected on the floor.

When they were done, they were shoved down the stairs and out into the courtyard. Around the corner of the building, Tukacier spotted something white heaped on the ground. It was "a pile of naked and almost naked bodies," the ten men and women who'd been thrown out of the window and whose corpses hadn't yet been collected. As Tukacier stared at the corpses, Cukurs appeared, striding toward him. Tukacier braced for a gunshot, but Cukurs only shouted for the Jews to look away. The aviator then turned to his underlings and spoke in a low voice. Tukacier heard one of the commandoes reply, "It's not so terrible, soon they'll also be part of this pile." The aviator alone among the commandoes seemed concerned about witnesses.

Tukacier and the others stood with their backs to the corpses. Minutes passed. Two of the prisoners, perhaps looking for loved ones, turned quickly to glance back at the bodies. The bucking sound of a shotgun echoed in the courtyard. Both Jews fell dead.

After an hour, Tukacier was marched back into the large hall, where Cukurs had become almost manic with bloodlust. In short order, Tukacier watched as he "beat to death 10 to 15 people." Tukacier was ordered to pick up the naked bodies, slick with blood, and bring them to the window overlooking the courtyard, where he pushed them out over the sill onto the heap below.

The night devolved into a slow-motion slaughter. The work was physically exhausting; the Jews didn't die easily. Schnapps and *zakuski*—hot and cold hors d'oeuvres of cured herring, hard cheeses, and small open-faced sandwiches—were brought for the Latvians. They would drink and snack, then return to clubbing the men and women. When dawn came, only about forty prisoners remained alive. One of the guards called out to them, asking who wanted to volunteer for a work detail. Tukacier raised his hand. He was taken to the Ministry of the Interior building, which he and some others were ordered to clean. The assignment allowed him to escape 19 Waldemars.

One night, Ella Medalje stood in the cellar listening to the sounds of revelry above. Soon after spotting Cukurs roaming the streets like a hunter, she'd been snatched up in a raid and brought to 19 Waldemars; now she waited to learn her fate. Around her were young Jewish women who'd been collected earlier in the day. Some of them moaned softly with pain; Cukurs' men had ripped off their earrings for their gold and gemstones, leaving the women's earlobes torn and bleeding. "After a few hours, the door opened suddenly and a drunk policeman appeared, a flashlight in one hand and a gun in the other," Ella recalled. The women lay prone on the floor as the beam of the flashlight flicked left and right; some of them drew their coats over their heads to hide their faces. Finally, the glow settled on the form of one girl sitting with her parents on the floor. The policeman stepped across the bodies and put a gun to her head. "You there, quickly upstairs!" he yelled. The girl and her family begged him not to take her. "Stop that noise!" he shouted hoarsely. "If you hesitate, you'll all be killed, you whores. We'll try all of you!" The young woman was taken, along with four or five others, only one of whom, a girl named Bavilska, Ella knew. The girls didn't return.

The next day, Ella was ordered upstairs to clean Cukurs' study. She dusted the furniture and tidied the papers and other things. At one point, she got down on her knees to clean under the sofa. She spotted something long and thin lying there, as if it had fallen or been shoved underneath and forgotten. Ella pulled it out. It was Bavilska's belt.

She returned to the basement. After many hours, guards came down and forced Ella and three remaining girls upstairs to the kitchen, where they were ordered to prepare breakfast. "What irony it was that we were to cook for the murderers of our relatives and friends," she said. "The murderers were Arājs, Cukurs . . . and the other Pērkonkrusts thugs. I will always remember their vile, loathsome, smug faces." The young women prepared the meal to the sounds of celebration on the floor above. After some time had passed, Ella turned and saw Cukurs enter the kitchen; he was standing in front of her, studying her face. "You don't look Jewish at all," he said finally. Ella didn't answer. Cukurs lingered in the kitchen; it was clear to Ella as she nervously went about preparing the food that the aviator saw himself as different from the other men who were drinking and cavorting upstairs. He was a world-famous aviator, after all, not just some inebriated frat boy (the Arājs Commando had recruited from some of the top fraternities) or an unemployed ditchdigger. "In those rare hours when he wasn't drunk, he liked to sport an air of elegance," Ella remembered. "He was always interested in meeting any new prisoners who could speak French." Ever since he'd landed in Poitiers during his epic flight to The Gambia and the French aviators had treated him with such kindness, the flier had had a soft spot for the country. When he found a French-speaking girl, he would talk to her for hours as she fried eggs and toasted bread. "Unfortunately," Ella noted, "this didn't stop him from sending her to the common grave with the others."

One afternoon, while Zelma was with her family in their tiny apartment, they heard a man calling up in German from the courtyard below: "Jewish girls, come out!" Zelma and her sister had no choice but to obey. They went down the stairs with the other girls and into the courtyard. Standing on the cobblestones was a German officer who'd come to choose a servant for his home. The young women formed two lines, and the officer walked between them, studying each one in turn. "I felt intuitively, he's going to pick me out," Zelma said. As so often happened in her life, her beauty set her apart from the other young women.

The officer stopped in front of her and said, "You are the one." Zelma, oddly, felt no fear. "I don't know why. He seemed to be a very peaceful man and I was right." Every morning, she went to the officer's apartment in a beautiful Riga building — the Germans, even more so than the Latvians, had their pick of the best flats — and did some light cleaning. After a while, she was transferred to another German officer, a man named Irving Hankelman. Her luck only improved. "He was a wonderful person." Zelma was shocked to find that Hankelman despised Hitler and regularly told her that the Third Reich was destined to self-destruct. "It's a terrible regime," he said. "This can't go on very long. He is desperate." Anyone caught talking like that, even a German officer, could be executed, but Hankelman seemed to trust Zelma. She sensed he even disapproved of the anti-Jewish laws. When he needed to write a pass for her, he'd put *"Fräulein Zelma"* on it. She would study the paper, then hand it back and remind him: "Mr. Hankelman, it's *die Judishe Fräulein.*" The officer would blanch. "I can't write it," he would say.

Every day, she went to Hankelman's apartment and cleaned and made lunch. One morning, however, she was picked up for work and instead of going to the officer's flat, she was taken to 19 Waldemars. Hauled out of the truck, she was brought to the basement. "I was terribly frightened," she said. "We had never heard of anyone returning from Pērkonkrusts." But instead of being brought upstairs, she was taken out on a work detail, moving heavy stones from buildings demolished during the German bombardment. Lifting and carrying the stones from one spot to the other was nonsense work meant to humiliate the Jewish workers. It was high summer and the heat was unrelenting. While Zelma was hauling the rocks, another woman, an elderly Jewish dentist, fainted and fell to the ground. Zelma and the other workers were upset that the frail woman had been forced to do such difficult work. One of the Latvian guards came over to Zelma and said, "Oh, you are not Jewish. You're making fun. You can get a Latvian passport." The guard was flirting with her. Zelma, already furious at the treatment of the dentist, felt her blood pound in her temples. "If this old lady can carry stones in the hot July sun, so can I," she shot back. "I am 100 percent Jewish. My father is a Jew and my mother is." She turned back to the stones.

Another time she was wiping down the windows at 19 Waldemars when she saw a group of Jews from the rich suburbs of the city being dragged to the courtyard. Through the windowpane, she watched as they were beaten to death. Their screams vibrated through the glass, "terrible shrieks and yelling." When a Jew fainted from the pain, a guard fetched a pail of cold water and splashed it over her head, then dropped the pail and began beating her again. Among the guards, Zelma recognized one of her former teachers.

Zelma embarked on a series of jobs dictated by the Germans: cleaning homes, removing rubble, working in various factories and kitchens. She spent her days with her work unit, then marched back to the ghetto to care for her parents. (Most Jews were allowed to return home to their families in the evening.) But one day she was brought to 19 Waldemars and pushed through the gate into the courtyard along with a group of other young women. There was a table in the center of the courtyard, and behind it sat a Latvian, quite possibly Cukurs. Police officers rushed out of the building to inspect the fresh crop of girls, who clustered together, heads bowed. The guards stole what they could from the terrified girls, ripping bracelets off wrists, rings off fingers, and earrings off earlobes. The Latvians seemed almost crazed; they ran around in a kind of delirium. Zelma spotted one she knew whose last name was Trucis, which sounds like the Latvian word for "rabbit." "I had always thought rabbits were lovely animals, very friendly," she said later. But this man was raving, lashing out at the girls. "I took him for a wild beast." The girls registered at the table and were driven toward a door in the building, then down some stairs into the dark basement, lit by only a small lamp in the ceiling. It was hot, stifling. There was no water and just one toilet in the corner. A single guard watched over them.

The selection process began. One after the other, they were brought upstairs when the Latvians called out their names. No one came back down. After some time, the guard called Zelma's name. She stepped carefully around the other girls and walked to the foot of the stairs. "Go up," the Latvian told her. She walked upstairs, into a room where the commandoes and the policemen were resting after their work. "They were all drunk, all had pistols in their hands." She was brought into another room where a man who appeared to be in his early thirties was waiting. He was trim and sharply handsome, with almost eerily symmetrical features in a long face; close-set eyes lent his gaze a hypnotic intensity. Zelma didn't know him. She looked around the room, her mind jumping feverishly from stratagem to stratagem; she was desperate to forestall what she felt was about to happen. But the door was probably guarded on the other side; if she ran, it most likely meant bludgeoning on the stairs and a slow death. "There was no way out."

The man came to her and began to grope her, mashing her flesh with his strong hands, yanking at the top of her dress. Zelma's defiance fled; she wept and begged the man for mercy. He seemed infuriated that she would even ask for such a thing, or perhaps he was angry that she didn't recognize him, didn't call him by his name. "You bitch," he shouted. "Don't you know who is standing before you? I am Viktor Arājs, the boss of this place."

After identifying himself, Arājs raped Zelma in a particularly violent way. By her own estimation, he "tortured" her. She screamed out and cried as he violated her; she lost track of time. The next thing she knew, she was being led back to the basement by a guard. The room was filled with the sound of weeping. Some of the girls' dresses had been torn open. There was no way to know whether the young women were crying about what had just happened or what they believed would happen next. She took her place on the floor.

After some time—Zelma couldn't tell whether it was during the night or the next day—the guard called out, "Who here is Zelma Shepshelovich?" She raised her hand. "It's me." She was brought into the courtyard and saw Irving Hankelman standing there. It turned out her nanny, Ieva, had followed the truck that had taken Zelma from their apartment. When Ieva saw the vehicle turn in to 19 Waldemars, she ran to the German officer and told him what had happened. Hankelman arrived at the building and demanded that Zelma be released. "He saved my life. He saved me from the Latvians," she said. It wasn't the first time and wouldn't be the last that a Jew or the friend of a Jew found themselves in the odd position of asking a Nazi officer for protection from the locals.

Zelma was ill for several days after she returned home. When she recovered, she inquired about the girls who'd been taken to 19 Waldemars with her. Their families informed her that none of their daughters had returned; only she had survived.

The city was slowly being drained of its Jews in batches of two or three hundred at a time. The house at 19 Waldemars would fill up and be emptied, fill up and be emptied again. Where had the Jews gone? Abram Shapiro, who now worked in the commandoes' garage, as well as playing the piano at his family's old apartment whenever Cukurs summoned him for one of his macabre parties, was one of the few in a position to report on what was happening. He told his friends that the commandoes' trucks went out in the morning loaded with clean spades and returned in the afternoon with "the shovels . . . smeared with blood and earth."

SIX

The Moscow Suburb

ON OCTOBER 24, THE GERMANS erected the last bit of barbed wire around a section of the Moscow suburb and declared the Riga ghetto closed. Still traumatized by her rape, Zelma returned to work—peeling potatoes in the kitchen of a home occupied by SS officers—to earn food money for her sick mother and father and her sister, Paula, who shared two small rooms in the ghetto with their relatives from Kuldīga. Each morning, she walked out of the gate guarded by Latvians and returned through it at night; each time she passed through, the Latvians searched her body and clothes for food and medicine. To and from her job, she was careful to wear her six-pointed yellow star.

The Germans strictly rationed food, allowing Jews only half the calories Latvians received. Sometimes Zelma's mother would mash potato peelings and make a pudding, often their only meal of the day. "There was a famine in the ghetto," she said. "My parents were always hungry." Every night when she returned home, she would hide an onion or a crust of bread stolen from her work in the hollow heel of her winter boot or in her "lion's mane" of auburn hair, which she wore up. The penalty for smuggling food into the ghetto was death, most often by a bullet to the head, but she couldn't stand to watch her parents growing frailer. The searching at the gate was hardly cursory — "they examined you *thoroughly,*" Zelma remembered — but day after day, the guards found nothing.

One evening when she arrived at the gate, a guard immediately pulled her out of the line. Someone had informed the SS officer in charge of the ghetto that she was smuggling in food. The German was waiting for her in his office; she must go to see him immediately. "Well, that's my end," she told her co-workers as she headed toward the SS building. "It was late summer," she recalled. "I was wearing, I remember, a nice dress." The man's office was on the second floor of the building; as she got closer, she could see him waiting for her on the landing, watching her approach. She climbed the stairs toward him.

When she stepped onto the landing, instead of ordering her into his office, the officer launched into a furious tirade. "We know that you are systematically taking food to the ghetto," he said, his voice loud and sharp. Zelma began to deny the accusation, but the German cut her off. "At 3 o'clock tomorrow some of our men are coming to the kitchen to pick you up. You're going to be shot."

It wasn't the officer's threats but his tone that struck her. He was shouting at her, a well-bred Jewish girl from Kuldīga, in public, as if she were some peasant without a proper upbringing. Something inside her rebelled; her face flushed with anger. "You may kill me if you want to," she said, "*but don't you dare* raise your voice at me." Even as the words escaped her, Zelma said later, she thought, "I don't believe myself I'm saying this."

The officer stared at her. "I think he was taken aback by what a little hungry Jewish girl would say to a high-ranking German officer," Zelma said. He didn't seem to know how to respond, so after a moment she turned and walked back down the stairs and made her way to her family's apartment. The next day, she sought out a friend at work and slipped into her hand a silver ring that had come from Palestine. "It was the dearest, most precious souvenir," she said. But she had no use for it any longer.

She said her farewells to her co-workers and waited for the soldiers to arrive. But three o'clock came and went, and no one showed up at the kitchen, no one asked her supervisors for the Shepshelovich girl. The day passed normally, and she went home that evening to her family, wondering whether her angry words had somehow saved her. Soon after, she was assigned to a new job: cleaning the five-room apartment of a German baron.

A month passed, and the coldest winter in living memory settled over Riga. On November 27, an announcement was posted on flyers pasted to houses and streetlamps: the ghetto was to be separated into two distinct areas. Four thousand able-bodied men were selected to be workers, while a much smaller number, about thirty, were chosen for medical experiments. The men were ordered to the northern part of the suburb, which would be known as the "small ghetto." The women and children would remain behind in the "large ghetto."

When it came time for the men to depart, Abram Shapiro gathered his mandolin, one of the few things from the happy prewar years he had managed to hold on to, along with some pictures of his father, now vanished and presumed dead. His mother took him aside. German soldiers and Latvian guards were shouting at the men to leave, pulling husbands from their wives and pushing boys back to their mothers. "You were saved the first time by standing next to me," she said amid the din. "Now you go with the men." He hugged her, then walked away with the older inmates.

On the morning of November 28, the Jews of the large ghetto awoke to find another set of notices pasted to walls on Sadovnikova Street. People came out of their houses and swarmed up and down the avenue, reading and rereading the decree. "The ghetto will be liquidated and its inmates will be evacuated," it read. "On Saturday, November 30th, all inmates must line up in closed columns of one thousand persons each." The western half of the large ghetto would be evacuated first. The eastern half would follow at a later date. "The decree hit the ghetto like a bolt of lightning," remembered one resident. "People stood stunned before the momentous announcement and kept trying to puzzle out the meaning of the words 'liquidated' and 'evacuated.'" Rumors, both bright and dark, began circulating immediately. Some Jews claimed that they'd seen barracks being built in the suburbs of Riga. Who else could they be for but Jewish workers? Others said that the evacuees would be taken to the train and then on to Poland, where the elderly would glue bags together and work would be found for the others.

Women began hurriedly knitting backpacks for the journey and stuffing them with clothes. Was it warm where they were going? Would clothes be provided to the workers? How was one to know? The smell of baking permeated the ghetto. Any ingredients that had been smuggled in past the Latvian guards were now brought out and made into bread or small cakes rather than left to go bad. People snuck off to the spots where they'd buried valuables and tried to remain hidden from the Latvian guards as they dug up their broaches and gold bracelets.

All day, there was a febrile mood in the apartments and on the streets. Something momentous was about to unfold, but what was it? There were about 25,000 Jews in the large ghetto, and there was no precedent for that many people being killed all at once. The mass extermination of European Jews hadn't yet begun. There were photos of Jews shot dead on Berlin's streets in the *Brown Book*, which had been passed around Riga, but no pictures of naked women being led to large pits. In fact, German forces to the east of Latvia were at that moment carrying out mass killings as they advanced toward Moscow, justified by Hitler's Barbarossa Decree, which authorized "ruthless and vigorous measures against Bolshevik inciters, guerrillas, saboteurs, Jews and the complete elimination of all active and passive resistance." But those atrocities were largely unknown in Latvia.

The temperature dipped below zero the night of the twenty-eighth;

snow fell heavily. As Zelma made her way home, she heard people shouting and crying in nearby streets. It seemed that panic had taken hold and Jews were frantically moving their household goods from one place to another, but in the confusion and the curtains of snow that partly obscured the scenes in front of her, she couldn't tell what was going on. What had happened? Had the evacuation been moved up? When she arrived home, her father was standing in the doorway. As soon as he spotted her, he asked her what she had seen in town. Zelma replied that there was nothing new to report. He nodded. "A very rich jeweler who took care of the apartments said those who go into town tomorrow will lose their hunger." Perhaps this was good news; perhaps they were to be moved to better quarters with sufficient food. Zelma assured her father that she was going to work as usual the next day.

But to her sister she said, "This is a bad sign. It means they are gathering the people in order to have them in one place." Zelma had decided that she wouldn't allow herself to be taken to the slaughter. "I'm not going to give them a chance to torture me, to rape me again and to kill me. I will die the way I find correct." She'd saved thirty-six pills — Ruminol, for treating anemia, and Veronal, a powerful sedative — and she intended to take them, then throw herself into the Daugava River, denying the Latvians and the Germans her death and her body. Paula didn't argue. "She was more clever than me," Zelma said, "but we had very similar characters." Once the Shepshelovich girls decided on something, there was almost nothing that could dissuade them.

The next morning, Paula, who usually stayed home while Zelma went to work, accompanied her sister on the walk to the gate. "I knew I had said goodbye to my parents for the last time," Zelma said, "and I knew I was saying goodbye to my sister." At the gate, she turned and kissed Paula. She waited for the work group to assemble — she was the only woman — and set off with them toward Alberta Street, then branched off to the German baron's apartment. The place was empty when she arrived, and she spent the day there, hidden behind a double door. She wrote a note to Ieva, her Latvian nanny. "I wanted her to know that I wasn't killed, that I didn't die after being raped." When evening came, she tore off the yellow star that had been sewn onto her coat; if she was caught by a German soldier, the punishment could be severe. She left the apartment and began walking to the building where Nank lived. She knew, without him saying anything, that he was beginning to care for her deeply; she felt he could be trusted to keep her secret. Zelma would give him the note, and he would deliver it to Ieva. Then she would take the pills, walk to the banks of the Daugava, and jump into the cold water.

When she got to Nank's apartment, she rang the doorbell. Her nerves were fraying; she felt "very upset" as she waited impatiently for him to open the door so that she could give him the note. After a moment, the Latvian, who appeared to be a bit inebriated, appeared at the door; behind him she could hear the sounds of a small party in progress. "I want you to deliver this note to my nanny," Zelma said, thrusting the paper at him. Nank sensed something was wrong. "Wait a moment," he said. "I'll come out. I don't want my friends to see you." He stepped onto the landing; Zelma handed him the note. He unfolded it and read it carefully. He looked at Zelma. "You're not going anywhere. I'm going to hide you."

Zelma, so intent on suicide a moment before, didn't protest. She allowed Nank to take her arm and lead her into the apartment.

In the Moscow suburb, the Jews waited in their homes, bundled against the cold that stole in through the gaps in the spavined wood. It was seven degrees below zero outside; the ground was frozen and covered by a "ringing, crackling frost." The sounds of the streets were muffled by seven inches of fallen snow. Few slept. For days, the ghetto had heaved and sparked with nervous energy, "in motion like ants in an anthill," said a woman named Greta Michelson. Now the backpacks were sitting by doors; the final selection of clothes and valuables had been made and the rest hidden or given away. The smell of cooking lingered.

Each family and each individual faced another in a series of binary choices: Was it better to take one's chances in the large ghetto or try to reach the small ghetto? Some said the latter was more dangerous, as the men had been the first to be killed when the Germans arrived in July. Wouldn't they logically continue the process and obliterate the small ghetto? But others said that surely the Germans would keep their best workers; if that was true, the men in the small ghetto stood a better chance of surviving. Some mothers made their decisions, found their husbands' left-behind suits and overcoats, and dressed their teenage sons to look like able-bodied workers, hoping to smuggle them to the men's barracks before the "evacuation" began. Other women scrounged or bargained for tattered men's clothes; they shed their dresses, donned the suits, and pulled their hair back under a hat or cap. Disguised as men, they hoped to make their way through the barbed wire into the workers' barracks. The young mothers among them secreted their infants beneath their coats. When they saw a gap in the patrols, they ran for the small ghetto.

At the same time, some teenage boys in the small ghetto had come to the opposite conclusion. Early that night, they waited by the doors of their houses, watching the guards pass by the barbed wire. Then they stepped out into the street, dashed over the fence, and ran back to their mothers in the large ghetto. It's possible to imagine members of both populations — those escaping *to* the small ghetto in hopes of saving themselves and those escaping *from* the small ghetto for the same reason — actually passing one another in the winter darkness.

Several married couples took large doses of morphine or pills they'd saved up and committed suicide together. One man told his wife that they should take their lives before the Germans and Latvians arrived. She refused. "Who will avenge us if we die?" she asked.

Abram Shapiro was in the small ghetto waiting with the other men, wondering whether his mother would succumb to the same fate as his father. There were no beds, so they slept on the floor and spent the night of November 29 listening. In the "Aryan" part of the city, Zelma was hiding in Nank's bedroom, waiting to hear the news of what the "evacuation" actually meant.

SEVEN

November 30

IN THE EARLY HOURS of the morning, the residents of the large ghetto began hearing the sound of vehicles approaching. Some of the Jews looked out their windows into the streets, where it was pitch-black; the sun wouldn't rise until after 8 a.m. There were needle-thin headlights along Lāčplēša and Jēkabpils Streets, and the black silhouettes of figures crossing in front of them. Then the muffled sound of fists banging on the front doors of buildings. "You have thirty minutes," a voice cried out in Latvian. "Be ready!" The voices — it was still difficult to see the men speaking — told the Jews to assemble on Sadovnikova Street with their luggage. The sleepless women roused their children, pulled on their coats, and lifted the backpacks onto their shoulders. At the opposite end of the ghetto, toward the east, soldiers could be seen cutting a hole in the barbed wire fence for a makeshift exit.

Through their windows, the Jews saw Latvian policemen and members of the Arājs Commando spreading out along the streets, the sight of their wool coats crisp against the white snow. Cukurs was there, directing the men, ordering them to their assignments. Some of them already "rolling drunk," the commandoes stomped up the stairs of the buildings, their footfalls softened by the deep snow, and went up into the stairwells. Moments later, there was a swell of human noise as they began driving the Jews out.

The first *aktion* (action), as the Germans called it, had started. Many cities and towns in Poland, Estonia, and Lithuania were witnessing similar violent roundups; the "Holocaust by bullets," which preceded the years of the concentration camps and gas chambers, began to unfold in occupied eastern Europe. The Germans used places like Riga and Vilnius as laboratories and proving grounds for the mass killing of Jews, employing *Einsatzgruppen* (literally, "task forces"), or paramilitary death squads, which worked alongside local militias such as the Arājs Commando.

Women and their children embraced as they waited for the Latvians to reach their floors. The first screams rose up above the sounds of knocking and furious male voices. Inside the buildings, some Jews were refusing to leave; others were too sick to get up from their beds. Roaring, the Latvians shouldered their way into the small rooms and lashed out with the stocks of their guns. They clubbed some women down, dragged others screaming into the stairwells. Those who resisted were shot in the head, their blood and brain matter spraying across the drab paint. Children were thrown down the stairwells as the hallways echoed with gunfire. After a building had been cleared, the Latvians went through the empty rooms searching for those who might be hiding. They probed the blankets and bunched-up sheets that had been left behind with the muzzles of their rifles. "Children were hidden among some mattresses and pillows," wrote one survivor, "and when they shook out the bedclothes . . . children, dead or wounded, dropped from them."

Most of the ghetto residents obeyed the orders. Columns of Jews stood silently on Sadovnikova Street, the steam of their breaths coiling upward into the predawn darkness. Apart from the soft footfalls of those joining the lines, it was still. Only the occasional cry of a woman's voice — "Oh, God!" — could be heard, followed quickly by a male voice shouting *"Klusu!"* (Quiet!). When a thousand had formed up in rows of five, they were led down the street toward the breach in the fence. One Latvian guard was in a frivolous mood. As the Jews streamed by, he called out, "Faster. Faster to Palestine!" The Jews in the eastern half of the large ghetto watched the columns pass from their windows, eager for clues to what the destination would be when their turn came to leave.

It was an ordinary Monday morning in the Aryan section of town, and the streets were beginning to be crowded with pedestrians on their way to work. Numbered electric streetcars, their long metal booms touching the latticework of wires above, swiveled around corners through the tiny banks of slush. Policemen ensconced in little round stations in the middle of intersections directed traffic.

The columns appeared. The Latvians stopped and stared as the poorly dressed Jews snaked through the city, the Latvian guards crying *"Ātrāk! Ātrāk!"* (Faster! Faster!). Some Jewish women, afraid of being shot, threw their backpacks down so they could make better time; the bundles grew into stacks as the morning progressed. But among others, a kind of terrified paralysis spread; more and more Jews were refusing to leave the ghetto or to advance down the road. "We are not going to Salaspils!" a voice shouted, referring to a concentration camp near Riga. "They are going to kill us. They are taking us to Rumbula." Rumbula was a large forest on the banks of the Daugava River, close to a major railway and a train station. Once through the fence, others succumbed to these premonitions and sat down on the macadam or broke ranks and fled with their children. The Latvians responded "unmercifully." Figures dashing from the columns were cut down by loud volleys of rifle fire. Children were bludgeoned into the snow and mothers shot in the face. The disobedience enraged Cukurs and his men, who'd been entrusted by the Germans with organizing a well-ordered evacuation. Now the Jews were throwing it into chaos.

One young girl marching along saw what she thought were cords of wood. But they were instead Jews who'd been shot and pulled to the side of the road, stacked one atop another by unknown hands. A woman found a Latvian on horseback and told him that she couldn't be resettled because she was due to go into the hospital for surgery that morning; he reached for his gun and shot her dead. "Already operated!" the man called out to his mates. The Riga nursing home, filled with elderly Jews, was "shot empty." On a road near the Škirotava station, a seven-year-old Latvian girl watched through the windows of her family's house as the stream of Jews tramped by a hundred feet away. Guards shoved the bundled-up figures along, shouting at them to pick up the pace. A bitter wailing filled the air. The girl's mother found her by the window and pulled her away, then hung a blanket over the glass.

Jews lay on the frost-rimed ground, gasping shallowly, some still alive though they'd been concussed or shot. Cukurs and the others went from body to body looking for those whose chests were rising and falling. When they steadied and then fired their revolvers, the victims' heads knocked gently against the cobblestones. A young Jewish man was walking on Ludzas Street near the ghetto when he saw Cukurs, dressed in black, commanding a squad of men who were clearing the streets of stragglers. "A Jewish woman started screaming when she was being dragged to the truck," the man said. "She wanted to have her daughter with her." Cukurs pulled his gun out of its holster and executed her. A few moments later, Cukurs came across a young boy calling out for his mother. "He killed this child with one shot."

Around 2 p.m., corpses began to clog the road, and the residents of the small ghetto were ordered out of their houses to clear them away. The men pulled the bodies from the road and loaded them on carts, surreptitiously checking for their loved ones. When they finished and were ordered to return to the small ghetto, one young man decided he wouldn't go back; instead, he hid in a yard behind a wooden gate. As he watched women and children being shoved along and beaten, he noticed a man in a uniform about fifteen steps away, his eyes scanning the road as if he was checking for stragglers. The man turned toward the gates and came closer to the straggler, who recognized him: "I realized in horror that it was Cukurs." He knew the aviator, or knew of him, and he went rigid with surprise. Cukurs looked up, and his eyes met the other man's. "I jumped away in fear," he remembered. Cukurs immediately raised his gun and fired at him, but missed.

As the men tramped back to their barracks, a policeman named Tuchel yelled for one of the Jews to stop. The man dashed away, and a shot cracked the air. The worker fell to the ground, dead. Another Jewish worker, fearing for his life, ran down Ludzas Street and hid. Some soldiers spotted him; among them he recognized Cukurs. The Latvian lifted his gun and was about to fire at the man, but he was cut off by a German officer, who swore at the worker, called him a pig, and ordered him to pull a nearby sledge heaped with corpses. The body of a baby swaddled in warm clothes was lying on the road. As Cukurs watched, the young man picked it up and placed it on top of the other corpses, then began to drag the sledge toward the Jewish cemetery, where a pit had been created using dynamite. On the way, he came across his mother's body lying on the street. He picked her up and carried her to the pit. At the lip of the trench, he propped her body on the edge, then let it fall. The Jews weren't allowed to say any prayers over their murdered loved ones. The body tumbled down onto a jumble of other corpses.

Many Jews were struck by one detail: the warm blood of the victims was seeping into the snow that had accumulated on the streets and melting it. The marchers were forced to walk over the red slush. "I remember it was squashing, *squish, squish,*" one survivor said. "Blood and snow, blood and snow." For days afterward, Jews would come across icicles of frozen blood branching out along sidewalks and into gutters.

The next morning, the men from the small ghetto lined up for roll call. As they marched out to their work assignments, they saw five hundred women lined up outside the barbed wire fence. The Germans had held them back to work as seamstresses. As he walked, Abram Shapiro spotted his mother; astonished, he rushed over the cobblestones to embrace her. "It was like a miracle," he said. The two cried as they held each other; his mother caressed his face and kissed him, until the Latvian guards came over and pushed the two apart.

No stragglers slipped back into the ghetto that night; Abram and the others waited for word of what had happened, but none arrived. Where had the women and children gone? Were they on their way to Poland, or were the rumors about the trenches among the stands of pines true?

When Abram returned that night from work, the women were out by the barbed wire again, and again Abram and his mother embraced. He memorized the small building where she was staying, and each morning and each evening, the two would find each other. It seemed strange to Abram that they'd been granted this reprieve, but it continued for a week. He and his mother would speak of each other's health and ordinary things. They would embrace, but after a few moments, the Latvian guards would begin separating the men and women. Abram and his mother would part with promises to see each other the next day.

Nank was true to his word; he kept Zelma safe in his apartment the night of the first *aktion*. He was, by this time, hopelessly in love with her. "He knew that, if I was found, it was his death and his family's death," Zelma said, but he never spoke about her leaving. Because the neighbors had known Zelma before the war and knew she was a Jew, he soon decided that they and his two roommates, Lidums and Kraujinš, needed to find another place to live. After they found a new flat and moved their possessions to it, Zelma hid in a small room that had a door with no knob. Nank kept the knob himself, so that no one could enter without his permission. "It was a beautiful small apartment," Zelma said. "Very cozy. One very large room shared by the others. Two rooms belonged to us." When Nank and his friends were out, she sat alone and listened to the BBC and Radio Moscow, the sound turned down low so the neighbors wouldn't hear.

Nank and Zelma became lovers. She wasn't in love with him, but he had protected her, and she felt that she owed him her life and that he wanted them to sleep together. (In fact, he wanted to marry her, but a wedding would expose her Jewish background.) Always thinking of her safety, Nank stole the passport of one of the Latvian girls who came to the apartment to sleep with his roommates, in case the Germans ever searched the house.

Did Zelma ever wonder if Nank would turn her in? News of such cases made the rounds among the Jews: The Latvian girl who'd just married her Jewish beau and was in the first months of pregnancy; not waiting for her husband's identity to be discovered, she "denounced him as a Jew" in front of German soldiers, who took him out and shot him. The Latvian farmer who hid a Jewish boy and his aunt in a tiny, dirt-walled room underneath the floorboards of his home while other fugitives were secreted in rooms above. Once they were inside, the two heard a lock turn. Panicked, they began to claw through the walls. Digging all night, they managed to tunnel out to the yard and run away into the fields. Days later, they learned that the farmer had turned the other Jews "hiding" in his house over to the authorities to be killed. "You can feel it, whether to trust them or not," the boy recalled years later.

After a few days of living with Lidums and Kraujinš, and listening to them talk about what they did each day, Zelma realized the two weren't bureaucrats like Nank but instead active members of the Arājs Commando. (Nank had agreed to share an apartment with them without knowing what they did, and on finding out, he didn't ask them to leave — perhaps that would have been awkward or dangerous.) After they had all been living together for a time, Nank told them that Zelma was Jewish and that if they informed on her, the Germans would question them about why they hadn't reported her sooner; their protests that they hadn't known she was Jewish, he said, would be dismissed outright. Lidums and Kraujinš knew he was right; ignorance was no excuse under German rule. They agreed to keep silent.

The apartment soon became a popular meeting place for the Arājs men, who arrived almost every evening after dark to drink vodka, eat, and wind down after an exhausting day. Zelma, who pretended to be just another Latvian girl, had been ushered into the very center of Cukurs' social circle. She and Nank greeted the men as they entered. "And please meet Zelma, my fiancée," Nank would say, and the commandoes would shake her hand, their hair slicked back with water, their boots streaked with mud. They smiled at her, courteous and respectful. They gathered in the living room and nodded approvingly as the vodka bottles came out. Sitting in wooden chairs and tippling their liquor, the soldiers seemed to recharge; they delighted in telling "stupid and naïve anecdotes," at which their friends laughed until they sputtered. There were certain rules for these gatherings. The commandoes' wives and girlfriends were never invited along, nor did the men talk about them, but "sometimes they brought the women they were sleeping with." The men, perhaps because of Zelma's presence, never spoke about women or sex. Even if one had hunted down a particularly appetizing Jewess that afternoon and ravished her in some revolting apartment, they kept it to themselves. Later, after the guests had left, she could hear those who were staying over, along with their prostitutes, on the bed in the room next to hers. "It was a terrible feeling . . . as if I were in a whorehouse," she recalled.

What did the Arājs men talk about during those long evenings? About war and death. "A story of the front line, a story of whom they had killed that night, how many they had killed," Zelma remembered. She watched them as they recited the details; she said little, only listened as the men described their days: the ridiculous attempt of this Jew or that Jew to evade their fate, running off until caught by a bullet in the neck and falling awkwardly, like a discarded puppet. There were so many comical ways the Jews died. Where did they think they were running to? Occasionally, one of the men would start singing, and the others would immediately join in, the sound of untrained male voices vibrating in the close air. It was as if they were replenishing themselves, readying their minds for the next mission.

"You can imagine what I felt like sitting among them," Zelma said, "but I had only one aim . . . remember, remember." She listened carefully as Nank greeted the commandoes; she memorized each name and any scrap of biographical information: what schools they went to, what jobs they held before the war, if they had belonged to a fraternity and which one, how many Jews they'd killed and on what dates. She didn't allow herself to react to the stories of Jews being shot; that would have been dangerous to both her and Nank. But she took in the dates, names, faces, the little anecdotes; she nodded every so often to show the men she was listening and she was with them in spirit. Perhaps she smiled to let them know that she was perfectly fine with hearing such stories. The images of the destruction of her friends and family arrived to her in these rough tales. Night after night the commandoes came, and she served them vodka, smiled at their compliments, and listened to them for hours. She might have been the first Jew to learn the complete truth about what was happening to her people.

At first, Zelma didn't understand why she was allowed to sit in the room while they were telling these stories, why Lidums and Kraujinš, the two members of the Arājs Commando, confided in her:

> There were evenings when [Nank] was out. They used to sit and talk to me and told me the truth. They told me terrible stories, unbelievable to a normal human being. Kraujinš told me a woman went to death with her baby. The woman couldn't go on with [it], it was a long way to the pit. So he took the baby out of her arms to keep it in his arms. It was such a nice baby, and the baby was counting the buttons of his great coat. And he said, "In order to make an end to it, I killed the baby myself."

The others might not have had any idea that she was Jewish, but Nank's two friends knew. Why make her a witness to what were, in fact, confessions? It was logical to assume, the way things were going, that very few Jews were going to survive the war, so perhaps they believed there was little risk in having her around. Or perhaps watching her listen to the stories was a kind of absolution. They'd killed a hundred Jews, but they'd kept Zelma safe another day. That had to mean something.

As the days passed, she sensed that something else motivated Lidums and Kraujinš, that these men were perhaps more cunning and more self-absorbed than she had given them credit for. The way they looked at her and spoke to her, it dawned on her that they seemed quite pleased with themselves. "They thought they had done me something great by not informing on me. They trusted me a lot. They thought, later on, that if the Germans lost the war, I will be the one who will defend them, who will forget what they have done." Zelma was their magical Jew.

One night, the apartment filled up with commandoes celebrating their day's work. The living room was crowded with male bodies; their boots tramped on the floor. Around midnight, Nank came to her and told her a special guest had just come in. "Herbert Cukurs had arrived in a leather coat, holding a pistol in his hand," she remembered. The men crowded around him. Cukurs addressed them; he showed them his famous gun. "With this pistol," he said, "I have killed 300 Jews today." The commandoes seemed duly impressed. It was the only time anyone would report hearing Cukurs confess to murder.

But in all the stories, Zelma never heard the names of her mother and father, or of her sister, Paula.

EIGHT

The Valley of the Dead

A WEEK AFTER THE FIRST *AKTION*, very early in the morning of December 8, this time without any warning, Cukurs and his men arrived in the eastern half of the large ghetto. They walked the streets, rapping on doors and calling out a half-hour warning. The remaining Jews, almost all women and children, were told to gather together and be ready to march. The roundups were taking place all over the country. In Liepāja, the last of the city's remaining eight hundred Jews were picked up and driven to the dunes of the nearby beach. "Papa, I'm afraid," one girl told her father as she walked toward the truck onto which the Jews were being loaded. "Don't be afraid, my child," her father answered. "It's Hanukah, a miracle's going to happen."

In Riga, Ella Medalje was working at a hospital, having escaped 19 Waldemars. That morning, SS units arrived in the wards and began searching for Jews, opening cupboards and inspecting supply rooms. Ella and the other women were ordered out of the hospital; in front, a truck was idling, with blue diesel smoke puffing from its tailpipe. She spotted a "healthy, tall" Latvian guard and decided to take a small risk. She walked in his direction, until she was standing next to the young man. "Where will you bring us?" she asked him quietly. The guard turned his head away and began to cry. "To the shooting," he said. "The whole morning I have been bringing our people there." The traces of hope of the remaining Jews were extinguished. "We were no longer people, only shadows," Ella later said. "Everything around us reminded us of butchery, spoke of another bloodbath to come."

In the predawn dark, piercing shouts could be heard from the rows of poor houses in the eastern section. The Jews were resisting. Some had barricaded themselves in their apartments, pushing furniture against the door or stacking suitcases there. Others sought out places to hide or tried to find a hole in the barbed wire fence through which to make a run. Cukurs and his men went from house to house, shooting or bludgeoning to death those who refused to line up into marching blocks. Outside the houses, some claimed later to have seen Cukurs taking babies from their mothers' arms and smashing their skulls against the pavement or walls. Inside the houses, the bodies piled up, blocking the stairwells. The death toll was much higher than on November 30; perhaps over one thousand bodies were left behind as the columns emerged from the Moscow suburb and trudged toward Rumbula.

Ella Medalje was taken by truck to the forest. When she was unloaded from the back, she found that the Latvian guards who'd cleared the ghetto were now formed up in a funnel leading somewhere she couldn't see. Cukurs would have been among them, his work in the ghetto done, though she didn't spot him. Naked Jews stood barefoot on the frozen ground as they waited for the Latvians to call them forward. "The people seemed indifferent to their surroundings," Ella said. "They were holding each other, suffering from the bitter cold, crying and saying goodbye." When the Latvians shouted at them to run and began hacking at them with their clubs, they let go of their loved ones and hurried down the funnel. Some were thrashed viciously when they hesitated, while others ran almost eagerly. From the forest ahead, an odd noise came through the trees, "a terrible hum, a drone."

The funnel narrowed as they went. At one point, a naked woman stopped and grabbed her grown son, who had somehow escaped the move to the small ghetto, and brought him over to a guard. She told the man that her son was a doctor and to kill him would be a waste. Surely the Germans could use him in one of their hospitals. He could save German soldiers! Before the guard could answer, the man turned to his mother and cried out, "Stop it!" He struggled away from her grasp and dashed toward the pits, where he was seen scrambling along the edge, neither hopping down into the trench nor trying to escape. The bystanders watched him run along the lip of the grave until a German sharpshooter brought up his rifle and fired a single shot. The young doctor disappeared into the pit.

A young woman named Frida Michelson, a dress designer, was running with a group of women. The first thing she saw was a German SS soldier standing next to a large wooden box; in one hand, he was gripping a wooden club. "Drop all your valuables and money in this box!" the man shouted over and over again as the Jews streamed past. The panicked women shoved the jewelry and money they'd hoarded for the trip into the slot and ran on. Next came a Latvian policeman who was crying, "Take off your coat and throw it on top of the rest." Frida began to panic. "My brain was working feverishly," she recalled. "The instinct for survival took hold of me." She took some documents — a passport and diplomas from several tailoring classes — from her coat pocket and ran up to the Latvian, spreading them out so he could read them. "I can bring lots of benefits to people," she said. "Look at my papers." The policeman smacked her hand, sending the papers fluttering away in the wind. "Go show your diplomas to Stalin!" he shouted at her. Dumbfounded, Frida shrugged her coat off her shoulders, tossed it on the tall pile, and, dressed only in a nightshirt, began running. From ahead of her, the droning sound grew sharper and more distinct, *tok, tok, TOK-TOK-TOK,* until she realized that it was composed of gunshots, one volley overtaken by another, the thick waves overlapping until it seemed that there could not be so many bullets in Riga.

The pits had been dug out by Russian prisoners over the course of three days and sloped into the earth like upside-down pyramids. A construction specialist had calculated the depth and size needed to fit up to 28,000 corpses and had given the measurements to the foremen. Now, on the rim, stood small groups of SS officers and some Latvians who were either killing off the wounded or just watching the action unfold. When they got to the edge of the pits, the Jews could see mounds of black and yellow sand piled high on either side. In the trench, naked corpses lay facedown, spread out over a hundred yards, their fish-pale skin in vivid contrast to the pitch-dark sand. (The Germans called this method of mass execution *Sardinenpackung,* or "sardine packing.") Here and there, the pile of bodies quivered as those who had been left alive below the top layer struggled to get air. When they entered the pits, the Jews were ordered to lie on the backs of those who'd just been shot, "still writhing and heaving, oozing blood, stinking of brains and excrement." A child old enough to walk was considered an adult and was lined up alongside the rest of their family. Only mothers with babies were exempted from this rule. They were ordered to hold their infants up in the air as they faced away from the gunmen. Then one soldier would shoot at the woman while another would take aim at her baby. Three pits were operating at one time.

Frida was close enough to the trenches to see a section of naked bodies and the shocks of black and red hair. "We were nearing the end," she said. "An indescribable fear took hold of me, a fear that bordered on loss of mind. I started screaming hysterically, tearing my hair, to drown out the sound of the shooting." Her screams merged with the rifle shots and the barks of *"Ātrāk! Ātrāk!"* (Faster! Faster!). The air buzzed weirdly with sound. Frida dashed toward the mountain of discarded clothes and threw herself into it, trying to burrow inside. A guard spotted her and slashed her across the back with his whip, yelling at her to stop playing games and to run faster. A policeman grabbed her and shouted obscenities in her face. Why was she still not undressed? He pushed her toward the trenches; nearby, another woman was protesting to a guard that she was a Latvian and had ended up in the procession of Jews by mistake. As the woman argued, Frida threw herself to the ground and pretended to be dead. "People were passing me, some stepped on me—I did not move."

An unearthly scream rose up behind her. Frida swiveled around. The guards had discovered another young woman in the pile of coats; like Frida, she had hidden there while the Latvians were distracted. "Some policemen rushed up to her and, like enraged animals, pounded her until she was dead." The body was left by the heap of woolen coats.

Time sped up, blurred, slowed down. Ella decided she would try to save her own life. She ran up to a Latvian she recognized from 19 Waldemars. The man was holding a whip, but his hands were shaking and he seemed to be suffering from an attack of nerves or sudden nausea. "He was very pale and could hardly stand on his feet." Ella plucked at his sleeve. "Save me," she said. "You know I am not Jewish."

The man seemed to be under a spell; he mumbled something unintelligible, then gestured toward a cluster of policemen. "Speak to the chiefs," he said. Ella hurried over to the men and immediately recognized Arājs, "his disfigured face resembling an animal's." She realized he was blind drunk, almost unable to stand. Jews were being pulled past him and his small cluster of commandoes toward the pits. Ella, shaking violently with fright and cold, stood in front of the unit's leader and said in a beseeching voice: "I am not Jewish." Arājs' eyes turned away from the stream of onrushing women and focused on her for a moment, then glanced past her. He waved with his hand. "Here everybody is Jewish," he said. "Today Jewish blood must flow!"

Ella shuffled away from the Latvians and found an SS officer to whom she repeated her claim that she had been brought to the pits by mistake. The German called to a Latvian, who took a coat from the pile and gave it to Ella, then told her to stand next to another supplicant, a young blond woman in her twenties. Ella knew the woman; she was Latvian but she'd gotten pregnant by her Jewish lover. Two weeks before, she'd given birth to a baby girl in the ghetto. She was holding the child in her arms as they waited.

The stream of walkers continued to flow in front of the pair. Without a word, the young mother thrust her infant into the arms of a woman passing by. The woman said nothing, only took the baby, held it to her naked chest, and kept walking down the line of guards on both sides. Ella and the mother stood together, watching as the woman moved farther away, until their view of her was blocked by the shoulders and heads of the women who followed in her wake. In a moment, she was gone.

As Frida Michelson lay on the freezing ground short of the trenches, something was dropped onto her back. Before she could think of what it could be, another hard object followed. She realized that Jews were throwing their shoes into a pile, as the guards ordered, and in their haste they had begun dropping them on what they thought was a corpse. "I was being covered with shoes, galoshes, felt boots. The load was heavy but I did not dare move a muscle." A woman ran past, saying "Ai! Ai! Ai!" and passed on. "I could hear people crying bitterly, parting with each other — and run, run, run." Lying at the bottom of the mountain of shoes, she listened as the number of shots slowly decreased, then petered out. She heard no more cries or moaning, but she did hear shovels digging into the dirt and voices speaking Russian.

By now, the sun was sinking in the west, and the last of the evacuees were led to the pits. Ella and the blond woman were ordered back to the road, where they saw a car with two other women sitting inside but no driver. Ella got in, as did the blond woman.

The driver's door was pulled open, and Herbert Cukurs lowered himself onto the seat. He flicked the switch for the interior light and turned around to look at the women. "For a few seconds, he stared at me," said Ella. "I remembered his assertion in the kitchen of their headquarters that I didn't look Jewish at all." After a moment, Cukurs turned and started the car. "He did not betray me," Ella said. "Perhaps he had seen too much blood that day, and he was now fed up with the constant shooting." Cukurs' eyes were unreadable. He might have been sickened by what he and his men had set in motion, or maybe killing Ella was just more work than he wanted at that moment.

Frida Michelson remained in the forest. She waited until the sound of the engines had faded, then dug herself out from underneath the shoes. In the quiet, she searched the immense pile of footwear that had built up and found a pair that fit her, then ran to a stand of pines and hid there.

As the night deepened and the temperature dipped, the fields and woods grew silent. Only a few Latvian sentries were left to guard the filled-in trenches; they smoked or chatted to pass the time. The pits, however, were not entirely quiet. Some of the Jews lying under the cover of sand had only been knocked unconscious by the bullets glancing off their skulls or cutting through the flesh of their necks. Others had pretended to be killed and were waiting for the Latvians and Germans to leave. Now they began to move, and the surface of the pits rippled like a sea under a light breeze. "Moans and whimpers" could be heard mixed in with the sound of the wind through the bare branches. Those who'd arrived early in the day, buried under layers of stiffening bodies, shouted for help. Hundreds of women and children in the lower depths of the pits suffocated. Here and there, the sand would shift or an eddy would suck the soil downward, and an arm or a face would appear. When this happened, a Latvian sentry would jog over the yellow sand, point his rifle, and shoot the survivor in the head, then shove the exposed body part back into the pit with the heel of his boot.

Abram Shapiro had woken that morning to the sound of gunfire. He and his friends left their bunks in the small ghetto, dressed quickly, and sought out the Jewish policemen who were permitted outside. "It's happening again," one told him, meaning another *aktion* was under way, one that might sweep up the seamstresses like his mother who had been previously spared. Abram returned to his bunk. He was overwhelmed by a feeling of powerlessness. He thought of embracing his mother a week before, and he felt a sensation of her physical being pass through his body. He lay in his bunk weeping.

After lining up for roll call, the men were ordered to clean the streets of corpses. Those chosen for the task hurried to the large ghetto. In the apartments there, they found dresses laid out or one shoe standing without its partner. The women had been surprised by the Latvians' arrival, and many had been pulled out into the streets with bare feet, others half-dressed or naked. There were meals on the tables, spoons resting in bowls of soup. Here and there, a wall sconce was still lit. Abram found no sign of his mother or sister; the house where they had been staying was empty.

The next morning, a small group of Latvian commuters was standing on the platform at the Rumbula station waiting for the train. Most likely among them were Riga residents heading out to the country to visit relatives or one of the small towns on business. When the wind shifted, the commuters were blanketed by a musty odor coming from the woods. Otherwise, the day was like any other weekday on the outskirts of Riga.

As they stood there in the chill, the Latvians saw two figures emerge out of the forest; from a distance, it looked as though they were wearing brown-and-white rags. When the two drew closer, the commuters realized it was a pair of naked women, their bodies streaked with mud. Blood was mixed in with the drying earth, and more blood, along with clumps of dirt, was smeared in their dark hair. One of the women had been shot in the neck, the other in the cheek and mouth. They came closer, gesturing for help with their hands; perhaps the one who'd been wounded in the neck also spoke, but no one recorded her words.

A Latvian soldier happened to be waiting for the train. He spotted the pair, raised his rifle, and pointed it at them. But the wife of a railway employee ran over, gesturing at the soldier and begging him to put down his gun. The two Jewish women stood mute, watching as the Latvian woman argued with the soldier. After a moment, it became clear that she wasn't asking the man to spare the women; she was simply pointing out that there were Latvian children nearby, watching, and it would be upsetting to execute the Jews in front of them. The guard lowered his gun, grabbed the women one by one, and lifted them onto a nearby cart. Then he turned the cart toward the pits and began taking them on their return journey.

The trucks vanished into the darkened streets of Riga, trundling toward the Arājs Commando garage, where the men climbed down from the cabs as the place filled with the smell of diesel fuel. Perhaps they washed up at their apartments or grabbed a quick dinner, but they wouldn't go to bed. Almost every night, there was a gathering, a party, and often it was at Nank's apartment, where Zelma Shepshelovich was impersonating a gentile girl.

On the night of December 8, Zelma felt restless; it was snowing outside, and Nank wasn't home. When her roommates returned home, they told her "that the street where I had lived with my parents, all the inhabitants had already been killed in Rumbula." A few Jews had escaped, however; perhaps others had been pulled away for work duty or for other inexplicable reasons. Zelma couldn't be sure her family was dead.

In the following days, Ieva brought to the apartment four or five Jews she'd found cleaning the courtyard of Nank's building. As they wolfed down some potatoes, Zelma recognized her old boyfriend, Max Schwartz, whom she'd expected to marry before the war broke out. She approached him and asked if he could find out who among their loved ones had survived. The next day, there was a knock at the door. Zelma opened it and saw Schwartz standing in the hallway. He looked at her and handed her a note before hurrying away. The message was written on cigarette paper; she opened it. "Dear Zelma," it read. "I'm happy you are in town. Your parents and your sister, along with my parents and my sister, were lined up for Rumbula on the night of December 8th and they were all shot dead." Schwartz had learned that Zelma's mother's last words were, "I'm happy Zelma is in town"—that is, that she was in hiding.

By that evening, in a country where Jews had been present since the year 1571, a place where not a single mention of a pogrom marred the written histories and to which Jews had fled for sanctuary to escape the eliminationist hatred that had lapped at its borders, about 60,000 Latvian Jews had been murdered in a little over five months. Riga was nearly empty of Jews, and the provinces hadn't been ignored either. Beginning that summer, Cukurs and the other commandoes had ventured out from the capital in blue Swedish-built municipal buses and private cars to towns and smaller cities across Latvia. When the convoy arrived, the commandoes would choose a local school building or farmhouse to serve as headquarters for their stay. Cukurs and the others would step off the bus and begin working right away. After gathering the Jews from the town or city, as well as tiny nearby hamlets and villages, the commandoes would line them up at the edge of a freshly dug pit and fire at their heads. Town after town, city after city, had been shot *Judenfrei*.

Survivors were few. Frida Michelson made it out of the forest and hid with a number of families in and around Riga, moving when the danger became too acute. She survived the war and eventually emigrated to Israel. Of Zelma's world, of her loved ones and neighbors and rabbis and teachers and crushes and the dozens and dozens of boys and girls she'd known at school—Orele, the brilliant neighbor who she'd gone dancing with, her high-minded college friends who'd sat in her tiny apartment for hours talking about opera and the latest French novels, the girl she'd given the Palestine ring to—almost nothing remained. Only Zelma and perhaps a few hundred others, perhaps a thousand in the whole country. First the Latvian Ieva had saved her, then the Nazi officer Hankelman, and now Nank, who'd made the decision to risk his life after seeing her once across a dance floor. She was protected daily by Lidums and Kraujinš, killers both. How could such love and vigilant care have been lavished on her while others were being slaughtered in the Valley of the Dead, as the Jews now called Rumbula?

When she was alone that night, she thought of the murdered thousands. "You have a mission," Zelma told herself. "Why did the rest perish, why did you stay alive?" She had endured the unthinkable but had survived, and now she sought to attribute some meaning to it. "I gave an oath to the dead," she said. The oath was clear and stern: "If you stay alive, you will destroy as many of them as you can."

PART TWO

Those Who Will Never Forget

NINE

A Latvian in Rio

In early 1942, Herbert Cukurs was spotted at the pits in the Bikernieki forest for the last time, firing a machine gun at a group of German Jews who'd been sent to Latvia for execution. Soon after, he and the other members of the Arājs Commando were redeployed to fight against partisan units in Belarus. The Jews of Latvia had been nearly exterminated, and new horizons awaited. They fought against Stalin's battalions but also murdered Gypsies, the mentally ill, and civilians; they left a string of burning villages in their wake. The aviator later published an account of a night attack on an enemy position. As in his articles years earlier publicizing his transcontinental flight, these scenes came alive in his writing; his report was an impressionistic, visceral evocation of battle. "The streaks of luminous bullets," he wrote, "like glittering yarns, cross the darkness of night; after short bursts of fire, the silence of the summer night takes over all around." By now, he'd taken on the vocabulary of Nazi propagandists, writing about the "sinking Jewish paradise" — that is, Russia — that was being crushed by the "New Europe," which was code for a Nazified Reich in the east.

Cukurs departed the Arājs Commando at the end of 1942; the reasons for his leaving remain unclear. After 1943, his movements become harder to trace. Some sources say that in February of that year, he enlisted in the Latvian Legion, a unit of the German Waffen-SS (the military branch of the SS), where he served as an "air observer" in battles with the Soviets. Other reports say that he returned to Latvia and settled in with his family at his estate in Bukaiši. As the Red Army fought its way toward the Latvian border, Cukurs took his wife and children and made his way to Germany, where he reportedly worked as an aviation technician for the firm Bücker Flugzeugbau. In the early days of 1945, the Soviets pressed the exhausted Wehrmacht, racing across Poland toward the Oder River, the last defensible natural barrier between the Red Army and Berlin. Cukurs gathered his family and fled to the forest near the German city of Kassel, where he stayed for several months. After the family emerged from the woods, Cukurs was either caught by forward elements of the US Army or surrendered to them; in either case, he was quickly released. His name didn't appear on the Americans' list of individuals to be detained, so, in the chaos of the postwar period, Cukurs went free.

The family made their way to Barcelona; Cukurs apparently sold some of the valuables he'd stolen from his Jewish victims in Riga to pay for the trip. There, Cukurs applied to the Catholic Church for help in obtaining a Brazilian visa, then moved on to Marseilles. The aviator was nearly broke by now but managed to find work in an airplane factory. When his visa finally came through, Cukurs spent the last of his money for tickets on the passenger ship *Cape of Good Hope,* which was sailing for Brazil. The ship docked at Rio de Janeiro on March 4, 1946, and Cukurs presented himself to the authorities as a political refugee who'd been persecuted by the Soviet communists. He received an identification card, No. 217180, which would allow him and his family to stay in Brazil for the foreseeable future.

"Our life in Brazil was difficult at first," remembered his daughter, Antinea, "as we had neither money, nor knew the language." To shelter his family, Cukurs, still vital at forty-six, built a primitive beach hut on a Rio beach from scavenged planks; this was later replaced by a floating boathouse. The Cukurses were squatters, but the paterfamilias had big plans and worked tirelessly to see them through. He sold his Leica camera and put the money into constructing pedal boats from scratch; he opened a boat rental company on the beach and watched it thrive. Cukurs moved on to speedboats, which he rented out to water-skiers and pleasure seekers. When this business took off, he found the money and the time to build an airplane from scratch, just as he had as a young man in Liepāja, and began taking sightseers up for views of Copacabana beach and the Christ the Redeemer statue atop Mount Corcovado. "Whatever Cukurs did," his daughter remembered, "became a sensation."

Unlike so many other war criminals, Cukurs arrived in his chosen country under his own name. Not only that, but he almost immediately began speaking out about his experiences during the war, which was, to say the least, highly unusual. He told anyone he could find that the Bolsheviks had exiled him from his native land and he was afraid to return. Days after arriving in Brazil, he actually sought out members of the local Jewish community and introduced himself around. The gregarious and charismatic Latvian managed to meet and charm merchants, doctors, lawyers, even the leaders of prominent Jewish organizations, who listened to his tales of escape and oppression with wonder.

Even this wasn't enough for Cukurs, however. He began to promote himself not only as a friend of the Jews in Rio but as a rescuer of them during the war. He told anyone who would listen that he'd hid Jews during the Nazi occupation and even delivered a number of them from execution. In fact, one of them, a young woman named Miriam Kaicners, had boarded the *Cape of Good Hope* with the Cukurs family. Soon after the ship docked, word spread that a Christian family had come to Rio with a Jewish girl they'd rescued from Hitler's troops. The Jews of Rio flocked to meet her.

"The Nazis discovered her in Lithuania," Cukurs told his new Jewish friends as he showed Miriam off at dinners and parties organized for his benefit. "She was marked for certain death. I saved her at the risk of my life." Miriam confirmed that Cukurs had hid her, brought her food, and provided her with false documents. The aviator's daughter, Antinea, recalled the night Miriam came to their estate in Bukaiši: "One evening I had already gone to bed when mother came into my room and a young woman along with her. Mother said my aunt had come and would sleep in my room. So she did for some time. The woman often wept, showed me the photos of her parents and sister — she said they were all dead. At times she would sing, then weep again . . . Even as a child, I understood that she was very sad."

Cukurs' listeners were moved by the story. After so many stories of terror and betrayal, here was a righteous Christian whose acts of bravery were confirmed by a Jew. Cukurs was welcomed into people's homes, thanked vociferously, and, if the aviator was to be believed, given money by grateful Jews to get on his feet in Brazil. Soon after his arrival, Cukurs was already "very popular" in Jewish circles. "They showed great respect for the brave Gentile who had saved a Jewish woman from the gas chambers," wrote one author.

Cukurs seemed unable to stop. He craved attention; he wished to be exalted. Some might see his actions in Rio as a cynical attempt to get his story out before any competing narratives emerged, or simply as a secret taunt directed at gullible Jews. But though he was many things, the Cukurs we find in his writings was rarely cynical. Something else was at work.

There were few, if any, Latvian Jews in Rio at the time, and Cukurs ran into no one who knew him from Riga. But other eyes were on the aviator. In the capitals of Europe, dozens of committees and organizations had sprung up after the war to help displaced Jews with jobs, housing, and resettlement in Palestine. These bureaus often took on grandiloquent titles and boasted imposing addresses in London or Prague or Vienna, though their offices were in many cases unheated one-room affairs staffed by a few undernourished refugees who worked tirelessly for no money and could barely afford the card stock their letters were printed on. These men and women had contacts worldwide, Jews from the same hometowns in Lithuania or Poland or Byelorussia who had sent an address from the far-flung cities they'd found themselves in after the war. Most were social service organizations, but a few, created by driven, often hyperactive men, burrowed deep into the cases of escaped Nazis. One name on these committees' radar was Herbert Cukurs.

On May 23, 1946, the Public Relations Sub-Committee of the Association of Baltic Jews in Great Britain issued a notice to its members. "The SS Obersturmführer Cukurs" — the term means "senior assault leader" and was not accurate in Cukurs' case — "is actually in . . . Rio de Janeiro . . . He was one of the executioners of Jews in Latvia . . . Taking advantage of the fact that there are no Jews of Latvian origin in Rio, he plays the part of a benefactor to people oppressed during the Nazi regime." The sub-committee issued a call for information on the aviator; before he could be brought to justice, his identity and exact location had to be confirmed beyond the shadow of a doubt. Letters went out to associates in Brazil, and telegrams crossed the Atlantic. ACCORDING STATEMENT RESPONSIBLE REFUGEES ARRIVED HERE FROM LETTLAND, read one, HERBERT CUKURS LEADER GHETTO RIGA TORTURED SADISTICALLY CAMP INHABITANTS KILLED HIMSELF CHILDREN. The recipients were asked to check if the stories were true, that Cukurs had escaped "Lettland" — that is, Latvia — and was currently living in Brazil.

While these amateur detectives tried to track down Cukurs, another ad hoc organization, the Committee for the Investigation of Nazi Crimes in Baltic Countries, made up of former concentration camp inmates from Riga and Buchenwald and the successor to the recently disbanded Group of Baltic Survivors in Great Britain, wrote to the British government to inquire about its plans for the Latvian killer. After the war, Britain had assumed responsibility for prosecuting war criminals in the Baltics; Cukurs' case fell under their bailiwick. A British MP by the name of Baron Frederick Elwyn-Jones wrote the committee back. "There is no doubt," said Jones, "that the crimes committed in and near Riga were ghastly, and I have every sympathy with the wish of the survivors and the relatives of the victims to bring those responsible to justice . . . [But] I am informed that much of the evidence submitted concerning the actions of individuals . . . is in fact hearsay." He was right; a number of the Cukurs testimonies, in fact, were not eyewitness accounts. Some recounted wild events; one popular story that circulated in Jewish communities after the war referred to the time Cukurs had drowned twelve hundred Jews in a lake near Riga. No such atrocity was ever confirmed; in fact, such conjecture paled in comparison with reality. The committees needed precise, firsthand, notarized eyewitness testimonies before they could move ahead.

The Committee for the Investigation of Nazi Crimes went to work. They asked their correspondents around the world to find Cukurs' victims and to get them to sign sworn affidavits about his crimes. And the affidavits were needed quickly, before Cukurs disappeared, as had so many other war criminals.

In a matter of weeks, statements came flowing in from New York, Tel Aviv, and points in between. Max Tukacier, who'd seen Cukurs savage Jews in the rooms of 19 Waldemars Street and shoot a young girl in the mouth, sent his. The orphaned musician Abram Shapiro, now known as Sasha Semenoff, mailed an affidavit from Las Vegas, where he was attempting to break into show business as a violinist and bandleader. After Riga, he'd been imprisoned in the Lenta concentration camp, where he'd caught scarlet fever and typhus; his weight had dropped to eighty pounds. On his way to the Stutthof concentration camp, a German soldier had demanded at gunpoint that he play the tune "La Paloma" on his mandolin. The song saved his life, and he kept the instrument with him everywhere he went after that. It became a totem of his survival and his ticket to America. In his testimony, Semenoff recounted the bloody shovels in the back of the commandoes' trucks and those awful nights at 4-4 Zaubes Street, when he played piano for the aviator and the other Latvian rapists, nights that he replayed endlessly in his mind.

Many eyewitnesses had perished; others had emigrated to new countries and were never heard from again. Zelma Shepshelovich was at that moment locked in an insane asylum in Riga; she'd written a long account of what she'd witnessed during the war, but it's unclear if it had reached the committee. But a half-dozen testimonies eventually arrived in the committee's mailbox. At least one was insufficiently official-looking for the members' taste; they sent peevish telegrams to the source, asking for notarized copies. Finally, after a great deal of detective work and money spent at Western Union, the dossier detailing the crimes of Herbert Cukurs was ready.

The next step was to prove the man rumored to be living in Brazil was actually the ex-commando. In early 1949, a Latvian Jew named Joseph Schneider slipped into the central library in Riga using the credentials of his brother, a college student studying aeronautics, and began searching through the aviation archive, combing the pages of the books and bound newspaper volumes for any mention of Cukurs. It didn't take long. Here were the glory days of "the Latvian Lindbergh" laid out in black and white, with photos from every angle and in a variety of attitudes: young, middle-aged, smiling, solemn, suited, uniformed, in a thick leather coat, sitting in a plane, shaking the hand of some notable in Indochina or another far-flung locale. Schneider had brought along a tiny razor blade, and while the archivist was otherwise occupied, he slipped the blade out of his pocket and positioned it between two of his fingers. Then he cut. "If there are today in Riga some damaged aviation books and volumes," Schneider said, "it is because of me." He slid the photos into his pocket, then returned the bound volumes with a nod and walked out of the library. He sent the pictures, as instructed, to the committee and a duplicate set to a contact in Rio with a brief cover letter, which began, "Here are the photographs of your uncle."

Half of the identification process was complete, but the Rio portion of the operation proved more difficult. Yet another organization, the Central Committee of Liberated Jews in the American Occupied Zone in Germany, wrote its contact in Brazil asking if they could get a photo of the man suspected to be Cukurs so that it could be compared to the ones taken from the Riga archive. A Jew known only as "Victor" volunteered to try. Along with an American friend, Victor arrived at the dock of the man they believed to be the Latvian killer and asked about renting a boat. The owner, a well-built middle-aged man, agreed. The three of them boarded the vessel, the owner turned the wheel toward the horizon and pushed the throttle up, and Victor began snapping photos of sailboats slipping over the water's fretted surface. After a while, he lazily turned the lens toward the boat's owner, who was wearing a battered captain's hat, and tried to snap a picture of him. But no matter how Victor maneuvered the camera, he couldn't get a decent shot. The captain turned this way and that, or looked away, or slipped down into the cockpit at the crucial moment. Victor snapped picture after picture of the back of the man's head or the top of his captain's hat. The man seemed terribly suspicious; he "always tried to hide his face," Victor reported. Even after an hour on the lake, the amateur agent wasn't convinced he'd gotten the shot.

When Victor developed the photos, they were indeed disappointing. He contacted the committee and volunteered to go back the next week for another try. "This time he will do it with a movie camera," a committee member wrote. "Let's hope the picture will be better." But that idea was apparently shelved; there's no record of any movie footage ever making it back to Europe. It turned out that Victor eventually found a shot of the captain that might be usable: in profile, ducked partly down inside the cockpit, with just his head and neck visible. It wasn't ideal; most of the photos that Joseph Schneider had collected in the Riga archives were face-on. But it would have to do. The Rio team sent the picture to the committee by airmail. And they waited.

After several weeks, word came back. The photos matched.

TEN

"The Epitome of Humanity"

BY 1950, HERBERT CUKURS was living a rich, happy life in Rio de Janeiro. His business was thriving, his children were bronzed and healthy, and he had friends "in all social sectors." Things were going so well that Cukurs' megalomania returned. He announced to friends that he wanted to start building airplane engines in Brazil for export; if some industrialist or banker could front him the money to construct a plant, he could start right away. Not only that, he was going to "create a great club and found a school of seaplane pilots unlike anything that had ever existed in Brazil." Perhaps it wasn't too late for a brilliant third act to his remarkably eventful life.

That spring, an editor at *O Cruzeiro,* Brazil's largest-circulation magazine, heard about the Latvian's heartwarming story. He visited Cukurs at his home in Rio and asked if he'd agree to an interview for a summer issue. *O Cruzeiro* was a pioneering, influential publication; every self-respecting middle-class family had the latest issue on their coffee table. If Cukurs valued his life and his freedom, he should have refused the offer.

He agreed immediately.

The piece hit newsstands and mailboxes on June 24, 1950, under the title "From the Baltics to Brazil," and it was nothing less than a paean to Herbert Cukurs as a man of destiny. The photo-essay was splashed over five pages, with pictures of the Latvian, still matinee-idol handsome, muscled, and golden brown in his dark nylon swimming trunks, cavorting on the crystal waters of Rio's lakes. It told the story of how the family had arrived in Brazil quite literally penniless on Carnival Sunday in 1946 and had slept on the beach until Herbert had his brainstorm for a boating empire. Now they were on their way to fame and fortune. "This is the story of a man who had to rebuild his entire existence, at forty-six years of age," the author wrote, "with the whole family, in a strange land of strange language, habits and climate." The point of the article was that the Cukurs were the best kind of people, and Brazil was lucky to have them. "When all seems lost, the human species can find within itself the same energies and ability enough to look forward, to forget the past and reshape its own destiny." The aviator's journey was "the epitome of humanity." The puff piece was accompanied by photographs of the smashingly good-looking family, who resembled the von Trapps relocated to a Brazilian beach.

Something deep and powerful, a kind of self-intoxication, was at work in Herbert Cukurs. What hunted war criminal would agree to pose for a five-page spread *under his own name* in the Brazilian equivalent of *Life* magazine, boasting of how good he had it? Most accused Nazis hid carefully, spent modestly, pruned their correspondents to a few. Obscurity was the price of survival. Cukurs did the opposite. It only added to the air of mystery and contradiction that surrounded his inner life.

The morning the June issue of *O Cruzeiro* was delivered to households across Brail, certain Jewish men and women in Rio and São Paulo paged slowly through the magazine as they ate their breakfast, browsing the article on how astrology had become the latest craze among American film stars or glancing at the ads for Kolynos dental cream or Hollywood cigarettes (*"Una tradiçáo do bom gusto!"*). Then they turned to page 23 and found themselves staring at a picture in the upper-left-hand corner of a still-young Herbert Cukurs, looking positively Olympian, basking in the South American sun. By now the name had become known in a few Jewish circles as belonging to a Jew-killer. Herbert Cukurs, the Butcher of Latvia! The monster in the black leather jacket! In *O Cruzeiro*!

The Jewish community immediately lobbied the Brazilian government to take action, and the Brazilians, in turn, sent a letter to London asking for information about the Cukurs case. The letter ended up on the desk of a British official named P. F. Hancock, who expressed his annoyance at the whole affair. "I am not quite happy about this," he wrote to a colleague. "The fact is that we are not prepared to pursue this business and we ought not to give the Brazilians the impression that we are . . . I am inclined to think that the right thing would be to do nothing at all." That is precisely what the British did. By 1950, London's interest in pursuing Nazi war criminals was waning. In the British government's view, the Nuremberg trials had dealt with the Nazi leaders most responsible for the Holocaust; the true masterminds were either dead or in jail, serving long sentences. The Cold War was under way, and West Germany was a key ally in the struggle against Stalin; pursuing SS killers and camp commandants could only serve to annoy the Brits' friends in Bonn. Foreign Office workers boxed up the materials they had collected on Cukurs, Viktor Arājs, and their compatriots and sent them to West Germany. Then they happily washed their hands of the whole affair.

Max Kaufmann, a Riga resident whose beloved only son had been murdered by Nazi officers near the end of the war, had learned years earlier that the ardor for catching escaped criminals was quickly fading in the West. After leaving Latvia, Kaufmann became consumed by the search for revenge and had spent the years after the war interviewing dozens and dozens of survivors for his massive book on the Latvian Holocaust, *Churbn Lettland.* He'd become a kind of one-man history department and global switchboard for Latvian Jews, able to tell his correspondents where their torturers had escaped to. When Kaufmann received a letter from a man in Brazil saying that the infamous Cukurs was living there, he wrote back in despair: "The matter of Cukurs has gone very near to my heart and is very well known to me. All you wrote is not new to me and I know all the details about him . . . But what good does it?! There is no reason to turn the matter over to the English or Americans, because they have no interest and no reason [to act on it]."

Kaufmann told the letter writer that he'd written to officials in São Paulo asking them to start proceedings against Cukurs, but "they did not even find it necessary to reply." The Jews of Rio were on their own.

To get to the bottom of Cukurs' story, a commission of leaders in the Jewish community met with Miriam Kaicners in Rio on August 14, 1950. How, they asked her, could she vouch for Herbert Cukurs, this deranged anti-Semitic killer? Miriam told them that she'd been a twenty-one-year-old student in Riga when the Germans invaded; she lost touch with her parents soon after and was arrested and brought to 19 Waldemars Street. The next morning, while she was sweeping the yard, Cukurs spotted the young woman and "ordered her to go to Zaubes Street in the fourth house, where he was residing." That would have been 4-4 Zaubes, the apartment of Abram Shapiro and his family. Cukurs kept her there for weeks, until he brought her to the house of a Jewish friend, Max Blumenau, and then on to his estate. The Jewish men pressed the young woman to explain why she'd stayed with the Latvian. "Although she admitted in some ways that Cukurs was a member of the SS and played a major role in the security police," the commission's report read, "she . . . knew nothing about the alleged atrocities." The meeting went on for hours, but Miriam was steadfast: "She claims that Cukurs has always been very good and correct with her and has saved her from the fate of thousands of Latvian Jews."

It appears that no one that day made the obvious connection to the young woman in Sasha Semenoff's original affidavit, the woman he spotted while playing the piano in his family's confiscated apartment, the girl whom multiple Arājs officers brutally raped. The time lines of the two stories match up, and Semenoff's assertion that Cukurs kept the Jewish girl in his flat for weeks aligns with Miriam's story of staying in the apartment for an extended time. It's likely, though not certain, that she was hiding from the commission — either because she was terrified of Cukurs or because she was ashamed — the fact that she had been raped in the apartment by her savior. It's not difficult to understand why Miriam had stayed with the aviator while in Latvia; leaving him would have meant almost certain death. But if she admitted that to the commission, they never spoke about it, and neither did she.

Rio's Jews decided to act themselves. They lobbied the Soviet consul in Brazil to extradite "*Cukurs der Judenmorder*" (Cukurs the Jew-killer) for trial; the Soviets refused, saying the nation of Latvia no longer existed and thus there were no grounds for the extradition. A hundred Jewish protestors stormed into Cukurs' marina and wrecked the place, smashing his pedal boats and scrawling graffiti on the walls accusing the aviator of war crimes. Three young Jewish men were arrested. On two other occasions, Cukurs was nearly run over by a car that seemed to swerve toward him on the street. The authorities detained one of the drivers, who was Jewish, but were unable to establish that he'd aimed his vehicle at Cukurs intentionally.

The city's newspapers ran wild; the story sucked up printer's ink by the drum. The aviator fought back against the charges, carefully constructing a story of a hero unjustly accused. His lawyer even compared his case to the famous Dreyfus affair in France, a provocative gesture, as Dreyfus was persecuted for being a Jew. Cukurs himself disputed the testimony of Sasha Semenoff and the others, and explained that during the initial 1941 roundups of Jews, he worked in the "road service," maintaining and cleaning the vehicles used by the Arājs Commando. At one point, he ran into Mrs. Shapiro, who begged him to help her son. "At the request of the young man's mother," he said, "I employed him in my service, so he would be protected." When the Germans tried moving the teenager to another assignment, away from Cukurs' protection, he put his foot down. "I said I could not do without the young man." After hearing what Semenoff said in his testimony, Cukurs claimed to be not only offended but mystified. "I saved him from death . . . He's alive because of me. And yet he files a deposition against me. Why? I don't know. Perhaps he was pressured." It was a smiling, brazen, madly cocksure performance, and large numbers of Brazilians were taken in by it. Would a guilty man really behave this way? they asked. Would he court publicity, advertise his role as a savior of Jewish women, challenge the survivors to look him squarely in the eye and prove their accusations? Surely, they said, the Jews had the wrong man.

Cukurs' victims and their families fumed. "When speaking of Cukurs, my father had nothing but a seething hate," said Sasha Semenoff's son, Paul, years later. "I am certain that if he helped my father in any way, it would have been part of his telling."

But Cukurs' behavior was so odd and his denials so fulsome that they prompted the questions: Did he somehow *think* he was innocent? Did he suffer from what would later be called a personality disorder, one that hid his own transgressions from his conscious mind? Or was he simply a sociopath? His wife, Milda, admitted that her husband did occasionally exhibit signs of intense anxiety. "He was a very kind person, incapable of harming anyone," she said, "although he was of a strong and extremely nervous character, which is why he shouted a lot, which made people think he was a violent man. But he instantly calmed down and embraced them affectionately, asking for forgiveness." In fact, he'd shown signs of violent behavior before the war. In March 1941, a Latvian newspaper published an account of how Cukurs had attacked his own son, beating him on the head and neck with the shaft of a whip. Could his denials indicate he was prone to dissociative episodes, in which his mind broke with reality? Or was he simply a man who felt his victims deserved what they got?

Brazil was a hotbed of psychoanalysis in the 1950s; the war had brought not only fugitive war criminals to South America but also Jewish analysts fleeing persecution in Berlin and Freud's Vienna. One

Rio newspaper editor consulted a specialist in criminal psychology and asked him to evaluate the Latvian's testimony. The psychologist sifted through the aviator's responses to the accusations and wrote up his results. "Cukurs . . . cannot disguise for long his familiarity and expertise in the Nazi technique of stereotyped defamation," the psychologist concluded. He was "a Nazi and authentically genocidal."

The story of the accused Nazi murderer spread to Europe, and a Latvian newspaper sent a reporter to interview the flier. In the published piece, Cukurs' distress is unmistakable. "I'm . . . drowning, with no end in sight," he told the reporter. Rash as ever, he even gave the newspaper his home address and asked them to print it. "If some compatriots want to come," he said, "please do." Under siege, he was in search of allies.

The aviator's livelihood was deeply affected by the controversy and the attacks on his marina. He moved to another neighborhood in Rio and opened a boat club. But the protestors found him, and the venture soon failed. Cukurs then moved on to Santos, fifty miles from São Paulo, and established Herbert & Children, an air taxi company. He'd dropped the "Cukurs," but the Jewish community discovered him anyway, and the business withered. Finally, he opened a modest boat rental business and air taxi company in the Interlagos neighborhood near São Paulo.

His family was pained by the accusations of wartime atrocities. "When I heard that my father was accused of such terrible things, I felt really bad," said his daughter, Antinea. "It hurt everybody. That day I had to go to school, and I told my father I didn't want to go." Cukurs gathered his children together — Gūnars, eighteen; Antinea, sixteen; and Herbert, eight — and talked about the charges. He told them about Latvia, what the war was like, and his part in it. And he categorically denied killing Jews. According to Antinea, he said, "'Even if I was a soldier and fought in World War I and II, this blood is not on my hands. My conscience is clean and calm.'" The stigma would follow Cukurs' children all their lives, and they stoutly defended him against all charges. "No one who flies can be a killer," his son Gūnars later said. "Aviators are idealists. My father was wronged like Jesus." Gūnars believed that the accusations resulted from jealousy over the success of his father's pedal boat business. After Cukurs refused an offer to buy him out, Gūnars told reporters, someone sprayed graffiti on their home calling him a Nazi. Then "Communist newspapers" fanned the flames.

As the controversy swirled, a memory from the war years came back to Antinea. When she was a young child, in 1942, the family was living on their estate in the country; she came down with a case of appendicitis. Cukurs, dressed in his aviator's uniform, put Antinea in his car and drove to Riga. Once they arrived, on the way to the hospital Antinea saw a tall man walking in the middle of the street. He was wearing striped clothes and a yellow star. She'd never seen such a curious outfit before. Just then, a horse-drawn carriage came racing by at high speed and nearly struck the man down. Terrified, Antinea cried out to her father: "Why is he walking on the road? Who is he?" Antinea recalled her father studying the figure. "That," he said, "is a very, very unhappy person." For Antinea, it was a sign that her father empathized with the Jews.

Cukurs was spooked. He went to the São Paulo police and asked for protection. He told them that "his life was endangered by Jews and he was afraid of being kidnapped." The police granted him a permit for a gun, and armed guards began patrolling in front of his house. The Department of Political and Social Order (DOPS), the Brazilian internal secret service widely feared for its torturing of dissidents and intellectuals, warned Jewish leaders in Rio that any attempt to kidnap Cukurs would meet with a serious response.

After a few months, the newspapers grew tired of Cukurs. The headlines stopped; the editorialists found other targets to obsess over. Jews in New York, Jerusalem, Vienna, London, Rio, and elsewhere had built a well-financed international crusade against the aviator, but they had little power to punish him. In 1960, the kidnapping of Adolf Eichmann in Argentina spurred a flurry of new reports on

Nazi criminals; Cukurs' name was among them. "From Riga, we send a warning," one Jewish correspondent wrote in a Brazilian newspaper three years after Eichmann was executed. "Herbert Cukurs! You will not escape! You will never be forgiven! Finally, justice will be done!" By any objective measure, however, the campaign against the Butcher of Latvia had been a miserable failure.

ELEVEN

Anton Kuenzle

IN PARIS, IN THE FALL OF 1964, the Mossad operative Jacob Medad, or Mio, began to lay the groundwork for the Cukurs operation. The testimonies he'd read in the slim folder in Yariv's apartment, which included that of Sasha Semenoff (formerly Abram Shapiro) and possibly that of Zelma Shepshelovich, had convinced him that killing Cukurs was necessary. According to another Mossad agent, "He really felt he was doing a holy mission. It was like closing a circuit for him."

In a way, Mio had been preparing for this operation since he was a teenager. He was in Palestine when the war broke out, studying at the Institute of Technology. "I remembered everything so vividly," he said, "as if it had only happened yesterday." His friends, some of whom, like him, still had loved ones in Berlin and Prague and Riga, said they were going to join the Haganah, the Jewish paramilitary organization. But the Haganah, apart from its Jewish Brigade Group, which would see action in North Africa and Italy, wasn't going to fire a single shot at the Germans; its soldiers would read about the war in the newspapers. Mio told his friends he had something else in mind; he was going to join the British army, which was actually fighting the Nazis. And that's what he did, becoming the first Jew in Palestine to sign up with the Brits. "It had already been said of me that I was a misfit," Mio said. Donning British khaki only proved it.

After serving with distinction in England, Mio returned to Palestine, joined the forces fighting for an independent Jewish homeland, and saw heavy action in the 1948 War of Independence. Tiring of military life, he left the army in 1955. That same year, he was approached by Mossad. The agency needed a dependable man to go to North Africa and ensure the safety of Jews living there. Would he do it? Mio agreed, and his career as an undercover operative began. He'd created and abandoned identities dozens and dozens of times since then, slipping into foreign countries with forged passports and escaping after the operation without leaving so much as a snapshot behind.

Now Mio's thoughts raced ahead to the intricacies of this mission. He'd already begun the transformation into Anton Kuenzle, who, he'd decided, should be a colorless, tough-minded, typically Austrian businessman. "I would become a different person," he said, "someone with an ordinary face that attracted no attention, who could disappear in any crowd without an effort." Mio seemed to relish the chance to become someone else.

Mossad told his family that they would be moving from their pleasant Parisian apartment to a new flat. If Mio was caught in Brazil, his wife and children could be targeted, so Mossad insisted on giving them a new address. Failure could bring other consequences, too. Mio might be executed by the right-wing Brazilian government, for starters. Conspiracists and anti-Semites around the globe would be in raptures at Israel's blunder; the Jewish state would be exposed to charges of cold-blooded murder. And the objective, a successful campaign against the German statute of limitations for prosecuting Nazi war crimes, which was already a long shot, would be smeared, perhaps fatally, with the spectacle of a failed plot on foreign soil. There would be little or no time for a second mission.

Mossad agents talk about something called "DAPA," which is an acronym for "plan B, plan C" in Hebrew. Every agent is supposed to prepare multiple options for escape if anything goes wrong in the field. Mio was famous within Mossad for being a true believer in the concept; later, when he taught at the academy, he would stress DAPA to new recruits. But the Cukurs operation would be different. He would be alone in Brazil, working with no lines of communication to Tel Aviv, no fallback identity, no fallback documents. There was literally no plan B. The first danger was exposure: Cukurs would suspect Mio and go to the police to have him detained. His cover was good enough to escape a cursory look by the Latvian, but it wouldn't hold up to a serious police investigation. The other danger was physical. "He's really going into the lion's lair," said one Mossad agent. "No backup team, no protection. If [Cukurs'] family is suspicious . . . , they can kill him, throw away his body. Nobody will know about it." Five months before, a coup d'état in Brazil had swept the elected president, João Goulart, out of office and ushered in a military dictatorship. The upheaval was still fresh, the country's politics turbulent and often menacing. Brazil was a police state whose authorities had made clear that they intended to protect Cukurs against any attempts to extradite him. The death penalty remained in effect. The same Israeli agent spelled out one scenario: "Can you imagine to yourself a Jewish Mossad agent being hanged because he tried to execute a Nazi?" It would not only be an embarrassment; it might also spur extremists in Brazil to attack Jews.

A solo mission naturally increased the mental strain on the operative. In fact, a mission in the Middle East that bore striking similarities to Mio's had recently ended badly. A young agent, a former army officer who'd emerged from Mossad's academy as a budding star, had been sent to Beirut on his first assignment. His psychological screenings were clean. "Everyone had great hopes for him as a man who would one day rise high in the service," wrote one historian of the agency. "He was keen and ambitious; he had proved himself in combat." Beirut was considered a good first posting: not as safe as London, but not as opaque or dangerous as, say, Baghdad.

The agent flew to Lebanon and checked into his hotel under his cover name. And then the strangest thing happened: the operative found he was unable to leave his room. The experience of handing his forged passport to the immigration officers at the airport had triggered a slow-moving nervous breakdown. Phone calls revealed a terrified man cowering behind his locked door. His superiors ordered the agent to leave the hotel and get to the airport, where he would board a plane for Israel, but he couldn't even make it to the elevator. It was a fiasco. "Mossad was faced," wrote one journalist, "with a bizarre situation in which it looked like one of its men might be spending the rest of his life in a smart Beirut hotel, a total recluse stranded forever far behind enemy lines." After debating other options, the agency finally called the CIA, which often worked closely with the Israelis. The Americans flew a team to Beirut and sent them to the agent's room, where they administered a powerful sedative to the Israeli, then spirited him out of the country on a stretcher. No one expected Mio to crack up in quite the same way, but this kind of mission was known to exert extreme pressure on a man's spirit.

The first thing Mio needed was travel documents. Mossad had two ways of obtaining false passports: theft and volunteers. The latter were called *sayanim*, Jews who'd spent time in foreign countries and were willing to help Israel in whatever way they could. A number of *sayanim* who returned to Israel were taken aside soon after their arrival and asked, "If when you're here in Israel we may need for security purposes to use your passport, would you allow us to do that?" Many said yes. There were other methods, too: fifty blank passports once vanished from a vault at the Canadian embassy in Vienna.

The agency maintained a fully stocked "travel department" in the basement of its Tel Aviv headquarters, employing highly trained forgers who manufactured visas and passports. Its vaults carried "every kind of ink," and the agency even made its own paper to match the different stocks used by various countries. Mossad devoted a great deal of money to this passport factory, and its specialists produced superior work; the documents were nearly indistinguishable from the real thing. The finished products ended up in a vault at Mossad headquarters, sorted by country. Within two days of the meeting with Yariv, Mio was handed an Austrian passport in the name of Anton Kuenzle.

On the morning of September 3, 1964, the operative made his way to the Gare du Nord train station in Paris and boarded the Trans Europ Express for Rotterdam, a port city where the locals barely batted an eye at the rich foreigners coming and going. As he'd never been to the Dutch city before, Mio ran little risk of stumbling on an old contact who'd want to know what he was doing there. He felt a rush of adrenaline as he approached the Dutch border; it was as if he were an actor on his way to the theater, eager to play a new role that embodied everything he wasn't. "In my daily life," he said later, "I was a quiet, introverted man, not particularly pushy or demanding . . . The minute I undertook a mission . . . I felt confident, even assertive, and had the capacity to strike up conversations and gain the trust of people I met." He felt far more assured in character than he did as himself. By the time the train pulled into the bustling Rotterdam station around noon, Mio was anxious to begin. He looked out at the crowd and saw everything from "sailors on brief leave to business people of varying degrees of respectability." It was a place that you could get lost in. Here he would create Anton Kuenzle.

The Hilton was booked, so he took a room at the upmarket Rijnhotel, left his suitcase on the bed, and asked for directions to the post office. There he met a "delightfully charming" clerk who helped him open a post office box. Mio presented his new Austrian passport, giving the hotel as his local address. He took the keys to the box and went looking for a bank. On Coolsingel Street, he spotted a branch of AMRO, well known throughout Europe; a man like Anton Kuenzle would do business with only the best firms. Thinking that the Dutch might still harbor some ill will toward the Germans, who'd occupied the country during the war, Mio switched to English when talking with the bank clerk. He produced $3,000 in American money, gave the post office box as his Rotterdam address, and within a few minutes was the owner of a checking account.

Mio quickly ticked off more items on his checklist: he booked a room at the Hilton for his return to Rotterdam in a week's time to wrap up his business, tipped the doorman handsomely to give him the basics on applying for a Brazilian visa, had a bellboy sent to the consulate to retrieve the necessary forms, and jumped in a taxi to get to his medical exam. He left the doctor's office with a certificate covered with "stamps of every possible shape and form," along with an international vaccination card. As he studied the visa forms, he realized that Brazil required three passport photos with the application. If the mission was successful, the photos would most likely be published in newspapers across the world. Mio needed to give Anton Kuenzle a new face.

The agent apparently never considered changing his hair; there wasn't much to work with, anyway — he was balding on top — and, besides, becoming a blond Austrian with black roots would be a giveaway. Hairpieces, too, were out — strictly for trashy Hollywood potboilers. He'd been growing a mustache since the meeting in Paris, and it was beginning to come in nicely. Now he decided he would add eyeglasses to the look. He walked the main business district until he found an optician's shop. Mio could have ordered eyeglasses with clear glass lenses that would have left his eyesight unaffected, but something Yosef Yariv, the mission commander, had mentioned about Cukurs had stayed with him. The Latvian, he'd said, was "cunning, mistrustful, ruthless and dangerous." The Eichmann abduction had set him on edge. What if, during some lunch in São Paulo, the aviator glanced at Mio's glasses from a certain angle and saw there was no curvature in the lenses? It wouldn't do.

Mio told the ophthalmologist that he'd been having trouble with his vision; the man stood him in front of the standard chart and Mio squinted. "I'm sorry, I don't seem to be able to read the bottom line," he said. The man wrote out a prescription and Mio picked out a dark, heavy frame. The glasses would be ready the next morning.

Next came stationery, envelopes, and business cards from a good printer's shop. Mio asked to proofread the materials once they were typeset, then set off to buy a new outfit. Before coming to Rotterdam, he'd asked around and learned that businessmen in Brazil tended to wear lightweight black suits. The tailor measured him and told him to come by the next day for a final fitting. The clothes would be ready in a week. As he went from shop to shop, he picked up "pocket matter": sales receipts, torn-off bus tickets, and a laundry stub. If Cukurs ever searched his room, the little items would lend his story authenticity. In his suitcase, he packed newly purchased Dutch soap and toothpaste with price tags from Rotterdam shops, in case Cukurs went as far as secretly inspecting his luggage.

The next day, Mio went back to the various shops he'd visited. When he slipped the glasses on at the optician's, he realized the prescription was more powerful than he'd thought; over time, the lenses might actually damage his eyes. But switching to clear glass would open a small crack in his own psyche that might betray him at a crucial moment. "My uncompromising perfectionism, even at the cost of my own health, is part of my nature," he said. "I always paid attention to all the small details, and I never took shortcuts." Only those who knew Mio well understood how deeply German he was.

To be a Berlin-born agent in Mossad wasn't a small thing. Mio didn't have the right kibbutz stories to tell; his pale skin didn't bronze like the native sabras'. Other citizens of the country, even other Mossad agents, often boycotted German goods and refused to set foot in a German car. When one agent took his family on a trip from France to Switzerland, they had to pass through Germany. His young son refused to step out of the car, even to pee. When the boy couldn't hold it any longer, the agent would pull over to the side of the Autobahn, and his son would stand up in the back seat and send a sparkling golden stream out the window and onto the grass that bordered the highway. This way, his feet never touched German soil. It was a gesture many Israelis would have understood.

Mio could have played down his past, but he never seriously considered it. While others boycotted German cars, he could be seen whipping around Tel Aviv in the latest Audi sedan. To relax, he did German crossword puzzles. "For Israel, he could give his life easily," his son said, "but deep inside, Germany didn't leave him." Mio didn't quite fit in at the agency, in part because the sabras exasperated him with their lusty seat-of-the-pants daring. "He didn't trust Israelis," his son said. "They didn't do things in an efficient, organized way like Germans." A sabra would have probably gotten the clear lenses. Not Mio.

After just two days in Rotterdam, Mio's business was done.

TWELVE

The Merciless One

AS MIO FLEW BACK TO Paris, a forty-two-year-old naturalized Israeli named Tuviah Friedman was boarding a plane in Tel Aviv on his way to a meeting that, if successful, could render the entire Cukurs operation pointless.

Friedman was one of the two leading Nazi-hunters in the world, second only to the more famous Viennese activist Simon Wiesenthal. Friedman's entire family had been murdered during the Holocaust, except for his father, who'd starved himself to death so that his children would have more to eat in the Polish ghetto. Friedman himself had barely escaped a sub-camp of Auschwitz by crawling on all fours through thick pools of human waste with four other inmates. Emaciated and vulnerable in the nearby woods, he and his friends were searching a field for buried potatoes when a local farmer spotted them. Friedman told his friends to stay where they were, then turned and walked toward the Pole. "Listen, you bastard!" he shouted as he approached. "Forget you ever saw us, understand? If you open your mouth . . . our men will burn your whole goddamned village down to the last stick of wood." Friedman was bluffing, but the farmer obeyed.

Very few of his brethren spoke to anyone like that in the winter of 1944. Like Mio, Tuviah Friedman was an unusual Jew.

After the war, Friedman took a position in a Polish intelligence unit in the city of Danzig, becoming one of the first official Jewish Nazi-hunters. At only twenty-three, he interrogated German soldiers, slowly sorting out the Nazi Party members from the regular Wehrmacht and delivering them to prison to await prosecution. He also went to investigate crime scenes. One day, he was asked to check out a number of buildings on the outskirts of Danzig where Nazis had reportedly held Jewish prisoners. He and the other investigators arrived at the first structure and opened the door. One large room was packed with the corpses of hundreds of naked men, women, and children, stacked to the ceiling. Another was "filled with boards on which were stretched human skins." The team spread out and found a building nearby whose door was secured with a heavy lock. They broke the lock, opened the door, and peered inside. There was an oven in the corner and lumps of something on the floor. They realized that the Germans had been using the room to melt down human fat cut from their victims to make soap. There were a few bars of it lying on the floor.

Friedman was tormented by these discoveries and by his interviews with the killers. But he took to the job; in fact, he grew to love it. When one afternoon Friedman left the jail after transferring the latest batch of prisoners, he realized he was as happy as he'd been in many months. "It may seem strange now, but as I walked home . . . I felt a tremendous sense of elation." He was punishing the guilty. The term "Nazi-hunter" is usually a metaphor. Nazi-hunters almost never physically touched or harmed the prey they pursued; in fact, they rarely met them. But Friedman was a hunter in the physical sense. He ferreted out SS and Gestapo men, brought them to his office, and beat them to extract information, beat them for their silences, beat them for their disrespect and the lies they told. When a man he knew to be an SS officer implicated in the murder of dozens of Jews swore that he'd been a simple private in the Wehrmacht and had never killed a single Yiddish speaker, Friedman would draw back his fist and smash it into the German's face. He would cut these men's lips, bloody their mouths, break their noses, if necessary; blood sprayed liberally onto his uniform, and he minded it not one bit. The other investigators began calling him "the merciless one." In the postwar period, he even hunted down and personally executed a few of the more notorious Nazis. "I lived for but one purpose," he said. "To find Nazi murderers, to beat them, and to have them removed from the world of decent people."

Why was Friedman so consumed by his mission? For him, it wasn't the idea that the Holocaust was being forgotten — as would soon become the common refrain of Jewish activists around the world — that motivated him. *It was that it had never been properly understood in the first place,* never comprehended in its fullness. The prevailing concept of the Holocaust in the 1940s and 1950s was the "guilty few" theory, which held that Hitler, Goebbels, and Himmler had cast a spell on the German people, manipulated them, victimized them even, into carrying out the Führer's personal vision of genocide. The Nuremberg trials, which targeted high-ranking Nazi leaders, reinforced this idea; it was the "great man" idea of history applied to the Shoah, and Friedman knew it to be wrong. He'd watched the ordinary killers at work, the sergeants and the corporals and the privates, in his Polish hometown of Radom; he'd watched Germans and Poles torturing Jews; he'd interviewed dozens and dozens of these perpetrators and knew they had been eager participants in the slaughter. It wasn't a few guilty ones who were responsible for the Holocaust; it was thousands of ordinary men who'd given themselves over to evil.

The world, in short, believed a lie. The justice it offered was pale and wholly unsatisfying. Many killers got just a few weeks for each Jewish "head" they sent into the pits. Friedman wasn't prepared to accept that.

In the city of Kovel, Poland, about 170 miles from Friedman's hometown, the Great Synagogue was used as a detention center for Jews rounded up by the Germans in the summer of 1942. Hundreds of men, women, and children were held there before German police squads and intelligence units ushered them out and shot them at the city's old Jewish cemetery. After the war, visitors found a hundred messages on the interior walls, written mostly in pencil. "My beloved Mama!" read one. "There was no escape. They brought us here from outside the ghetto, and now we must die a terrible death . . . We kiss you over and over." Another: "Forgive me! Mother I want you to know they caught me when I went to bring water. If you come here, remember your daughter, Yente Sofer, who was murdered on 14. 9. 1942." Tanya Arbeiter, trapped with her family, wrote on August 23: "Hush, the murderers are coming . . . Hearing their voices stops our hearts beating. Lord, take us to Your eternity!" Most of the notes were personal, written to specific loved ones, but more than one was addressed to all Jews as a tribe, and many of those had raw violence in them. "You who come after us—Remember!" Yehuda Schechter wrote. "We demand vengeance! Cruel vengeance!" Later, the walls were painted over and the synagogue turned into a textile factory. Before the refurbishing, local Jews copied down the inscriptions. One note, left by two young girls, read: "One so wants to live and they won't allow it. Revenge. Revenge."

It's unclear whether Friedman ever read the messages, but their tone resonated in his soul. The soon-to-be-murdered didn't ask for justice; that word doesn't appear once in the hundred inscriptions on the synagogue walls. But "vengeance" appears three times, "revenge" six. Friedman interpreted the wishes of the murdered not in the light of Judaic ideas of mercy, but precisely as they'd been written.

Later, Friedman moved to Haifa, Israel, with his wife, Anna, an eye surgeon. He gave up terrorizing suspects, but his obsession with escaped Nazis grew over the years to nearly unmanageable proportions. He worked for several years as director of the Haifa branch of Yad Vashem, Israel's memorial to the victims of the Holocaust, but was dismissed for prioritizing the pursuit of war criminals over research and preserving the memory of the dead. It's not easy to get fired from Yad Vashem for focusing on Nazis, but Friedman somehow managed it. In 1957, he founded the Institute for the Documentation of Nazi War Crimes, which the Israeli government, focused on present dangers such as the hostility of its Arab neighbors, begrudgingly supported with small yearly grants. The perpetrators of the Shoah were all Friedman seemed to be able to think about, or talk about, or work on. Every day, his "customers"—the survivors, some of them now growing stooped and gray—came into his little office in Haifa and spoke for hours about how their loved ones had been abducted or butchered in front of them. He spent his days writing down the places and the dates where these things had occurred, along with the names of the killers, or giving occasional press conferences, which were increasingly less well attended as the years went on. He had thousands of affidavits in his files pointing to the guilt of an equal number of Nazis, but when he offered them to German prosecutors, they declined. No one wanted to try the cases. Britain and America were focused on confronting the USSR. Germany had been paying heavy reparations to the new state of Israel since 1952, sparking a consumer boom that Friedman—who couldn't afford any of the new luxuries on his nonexistent salary—noticed on his way to and from work. "On all sides were new and valuable goods," he remembered. "Diesel engines, ships, consumer goods. It was so easy to look away, to forget, to enjoy these shiny things." The days when his efforts had produced a kind of euphoria were forgotten. He was entering his thirties now, balding and unprosperous. Many nights, he returned to the small flat he shared with Anna more depressed than when he'd left.

There was trouble there, too. "Things were not good at home," he admitted. He wasn't very pleasant company, for one thing, as almost the only thing he could talk about were Nazis. He didn't contribute to the couple's finances. The Friedmans were living in a cheap apartment in Haifa with practically no money; Friedman's bank account was "next to zero." The government had recently set its contribution to the institute at 200 Israeli pounds, not enough to pay for the rented typewriter, the phone bill, and the postage, which made drawing even a small salary impossible. "All right," Friedman thought when he got the news, "I will eat less. I will buy nothing."

People laughed at him. "He pestered and annoyed," wrote the Israeli newspaper *Maariv*. "He was looked upon as . . . an obsessed fanatic. Doors were closed in his face." One day in the 1950s, Friedman was walking from his office in Haifa when he heard someone calling to him from a passing bus. "Eichmann!" the man cried out. "Shalom, shalom! Eichmann!" Friedman stood still as the bus droned past and the sound of the engine rumbled away, but the almost joyful voice still rang in his ears. There could be no doubt whom the man was calling to. To be mocked by a fellow Israeli, to be called by the name of the evil one, cut Friedman to the heart.

One night, he began talking once again about Nazis. He never revealed what exactly the theme was that evening, perhaps a new lead that was particularly promising or a theory he'd been developing about the South American ratlines. Anna's voice suddenly cut him off.

"Enough!"

Friedman looked up at his wife in amazement. He'd never seen her so furious.

"It's enough, already," Anna said. "The war has been over thirteen years . . . and you are running around like a fanatic . . . All the big shots in Israel don't care, everybody wants to forget, and you, you think, Tadek, that a small man like you can go against everybody?" He thought he was helping his Jewish clients, but every time he gave a press conference or a newspaper ran a story about one of his quests to find this murderer or that murderer, Israelis were transported back to the morning they arrived at Treblinka, or to the day in Warsaw or some godforsaken Lithuanian town when their fathers were dragged out the door by the Fascists. He wasn't helping Jews, he was causing them excruciating pain, and he couldn't see it. "They want to forget," she said.

Friedman listened. Israelis mocked him; his wife thought he was hurting Jews. "I said nothing," he remembered. "I was thirty-seven years old, and defeated. I felt that my life was a mockery, even a farce . . . I had come to the end of the road. Ahead, a sheer, unscalable mountain confronted me. I was alone."

A month later, Anna suffered a miscarriage, perhaps due to the stress of their living situation. In the hospital, Friedman tried to comfort her. She told him, with great bitterness, that he had made her life miserable. He had pulled her into "this whirlpool of Nazism that was spinning us both about, draining our strength and destroying us." She told him she regretted marrying him.

Friedman let Anna finish and was quiet for a moment, then spoke. He volunteered to leave the apartment when she returned, so that she could be alone. He would go live on a kibbutz if that would ease the despair she felt. Divorce, separation, whatever she wanted. But under no circumstances, he told her, and not for a single day would he stop pursuing the perpetrators of the European Shoah. "I could not turn back." The marriage survived, but Anna remained deeply unhappy about her husband's obsession.

That was 1959. A year later, Friedman learned of the statute of limitations that would stop the prosecutions of unindicted Nazi murderers on May 8, 1965. He decided to drop all his cases against fugitive war criminals, the ones that filled the file cabinets at his institute, and devote his life to creating a campaign to block the amnesty. He canceled his plans to get a degree at the University of Zurich, something he'd long dreamt of doing, and cut his household budget even further. "I had come to the conclusion that it was useless for me to spend these last years searching for 20 or 30 Nazis, while the statute of limitations, if it were to go into effect, would permit hundreds of Nazis to come out of their hiding and again swagger into the beer cellars, beyond the arm of the law." Friedman had no illusions that the campaign would go smoothly. He sensed the leaders in Bonn, backed by the majority of voters, were desperate to end the Nazi trials and free Germany from its shame. They would resist his efforts. "Our fight with the West German government," he said, "was a desperate and bitter struggle."

Now, four years later, as he prepared to fly to Bonn to speak with the country's justice minister, Ewald Bucher, he appeared to be making progress. Bucher, a boyish-looking former Nazi, had become the government's point man on the statute issue. Friedman had brought along several files on the most egregious Nazi criminals still at large, which he intended to present to the minister. The details inside were so clear and so overwhelming that Friedman felt they would convince any fair-minded man that the perpetrators must never be allowed to go free. He was hopeful about his mission, a rather novel emotion for him. "I felt sure . . . I would be able to induce [Bucher] to agree to the extension or cancellation of the statute."

After landing in Zurich, a Jewish friend picked him up in a rented Chevrolet. The two crossed the border into Germany and flew along the Autobahn at high speed, making dark jokes about the paranoia they felt as Jews who'd ventured back to the source of Hitlerian evil. To dampen the anxiety, his friend smoked constantly, "crunching the cigarettes between his teeth." Friedman, too, felt spooked. He'd been in Germany the year before and visited the Sachsenhausen concentration camp; during the tour, the guide had told Friedman and his Jewish companions how the guards had hung prisoners by one foot and one arm and then fed the men, still alive, to their dogs as "a fine meal." One of his friends went pale and said, "Friedman, I feel ill, I can't stand it any longer. I feel as though I am in hell. Let's get out." Friedman experienced the same symptoms. "A strange fear had gripped me, as though once again the SS jackboots were lording it over their defenseless prisoners, and the end of the world was in sight."

Despite the memories, Friedman and his friend arrived in Bonn unscathed.

The next morning, Friedman met with Bucher and his assistant. As the justice minister listened, Friedman recounted the days of the *aktions* in his hometown. Telling the story, he felt his nerves twitch. "I myself seemed . . . back in the ghetto of Radom and was going through the horrors again." The assistant took out a handkerchief from his breast pocket and wiped away tears, while the minister shook his head and said softly, "Terrible, Herr Friedman, that is really *schrecklich.*" But Friedman, despite his excellent interrogation skills, couldn't tell if he was getting through to Bucher, who remained poised throughout the meeting. Though he was quick to pick up subtle hints in facial expressions and tone, Friedman felt unsure of himself. These "new" Germans, so polite, so bland, so sympathetic — were they real or not?

Friedman decided to change his tone. "We do not want any lynchings," he told Bucher calmly. The word must have been rather a shock to Bucher. *Lynchings?* Who had mentioned lynchings? "But we insist on trials and justice," Friedman continued, rather icily. "I cannot guarantee, Herr Minister, that if the Statute [goes] into force, that no acts of retribution will occur on your streets." The Nazi-hunter had issued what was quite clearly a threat. If Germany pardoned the killers, Jews might be forced to take action and execute the culprits one by one. In public, in Germany. It was unclear on whose authority Friedman proposed this rather controversial idea; he was in touch with leaders high in the Israeli government, including the minister of foreign affairs, Golda Meir, but there was no evidence they'd sanctioned talk of assassinations on Germany's streets. (It's also unclear whether Friedman knew about the Cukurs mission. Given the high degree of secrecy that Mossad maintained, it's highly unlikely.) If the warning was a bluff, it's an excellent one. The image of Nazi killers being shot down in the middle of Berlin, inflaming public opinion and stoking an unpredictable response from the German public, couldn't have been an appealing one for the justice minister.

Bucher was called away to the phone. His assistant leaned over to Friedman. "The Justice Minister himself is looking for a way of extending the limitation," he said, "although in public and to the press he maintains the opposite, in order to calm the former Nazis." Friedman was heartened; this was excellent news. If the justice minister, the public face of the pro-statute cause, was secretly on Friedman's side, it was a wonderful development. The minister finished his call and walked Friedman out of the office. Soon after, Bucher would announce a commission to investigate the government's response to Nazi crimes; he would also issue a call to all nations to search their archives and provide any materials on murders committed during the war to the West German government.

As he flew back to Tel Aviv, Friedman leaned back in his economy seat, freshly confident. He was relieved to be leaving Germany, but it had been a successful trip. Bucher and his assistant had been keenly sympathetic — at least he *thought* they had, it was difficult to tell — and the news of the minister's secret campaign against the statute was promising, to say the least. The Nazi-hunter now believed he "was winning my fight" with the Germans. "I felt that Dr. Bucher was a man on whom I could rely," he remembered. Perhaps soon he could at last matriculate at the University of Zurich and return to the hunt for Dr. Josef Mengele, "the Angel of Death," and others, secure in the knowledge that the monsters of the Shoah would still be hunted by the country that produced them.

THIRTEEN

The Late One

IN PARIS, MIO WENT AGAIN to the apartment on the Avenue de Versailles, where Yariv was waiting. There they reviewed the operation. Mio told the spy chief that Mossad should write him letters in invisible ink and send them to the American Express office in São Paulo or whichever Brazilian city he was staying in. He would write his reports in the same way, disguised as business letters. They also gave Cukurs a code name: "the Late One" (as in "the deceased"), a bit of Jewish gallows humor they would use exclusively when talking about Herbert Cukurs.

Other details had been hashed out earlier. Cukurs would be spirited out of Brazil to another South American country, where the final act would take place. Brazil was a military dictatorship that still enforced the death penalty; Mossad wanted to avoid the risk of having one or more of its agents publicly executed if the operation went bad. And the country was home to thousands of Jews, who would face blowback if the killing was carried out on native soil. A more liberal country — Uruguay, for instance — was preferable.

Mio flew to Zurich, a center of international banking whose luster would give his fake business instant credibility. He made his way to the Credit Suisse bank and deposited $6,000, then asked the clerk to send a notice to the bank's officers in Brazil that an Anton Kuenzle would soon be visiting their country and might be paying a call on them. The clerk agreed and handed Mio the letter of credit in Kuenzle's name. Back in Paris, it was time to take the photos for his Brazilian visa. Mio's mustache was still coming in, so he hired a professional makeup artist to thicken it, then visited a photographer's studio, where he stared solemnly at the camera. "Had I known then . . . that half a year later my photo would feature prominently on Interpol posters below the headline 'Wanted for Murder,' I might have allowed a friendlier . . . expression," he later said. He returned to Rotterdam, picked up his visa, his tropical-weight suit, and his freshly printed stationery, then took the train home to Paris. After kissing his wife and saying goodbye to their children, now ensconced in their new apartment, he boarded an Air France flight for Rio.

As he flew toward the New World under his new identity, Mio joined the select few Mossad agents working to find and kill former Nazis. There may have, in fact, been only one other undercover operative who shared the same mission in 1964. In Damascus, a thirty-nine-year-old exporter and firebrand Baath Party member named Kamel Amin Thaabet had made a deep impression since arriving in the country two years before. The flashy, brilliant Thaabet was a wealthy socialite with connections among generals, ministers, and other officials high in the government, including the president, Amin al-Hafiz. His name had even been bandied about to fill the empty slot as the next deputy minister of defense. His real name, however, wasn't Thaabet but Eli Cohen, and he was a Mossad agent working undercover for the agency. In his apartment, nothing was as it seemed: the aspirin bottle was actually filled with cyanide tablets; on the bathroom sink, hollowed-out bars of the fragrant and expensive Yardley soap hid plastic explosives; inside the high-end American blender was a miniature radio transmitter, with another secreted beneath the kitchen floor. The bathroom doubled as a photo lab.

Cohen had done some good work on escaped Nazis. Through a sheikh he'd befriended, he met a friendly German man named

Rosello who had dealt in "Jewish affairs" during the war. Rosello's real name turned out to be Franz Rademacher. For three years, from May 1940 until April 1943, he'd served in the German Foreign Office as the head of Referat D III, then called the Judenreferat, or "Jewish desk." In that role, Rademacher had facilitated the deaths of thousands; in one expense report he filed for a journey undertaken for D III, under "purpose of trip" he wrote "Liquidate the Jews."

Once Tel Aviv approved a mission to eliminate Rademacher, Cohen spent several nights at his dining room table mixing chemicals into a powder. He held the powder over a flame until it melted, then took a brush and carefully covered the inside of an ordinary business envelope with the explosive before placing a small detonator inside and sealing it up in a way that would avoid a premature blast. He addressed the letter to "Thom Rosello," walked to the nearby post office, and slipped it into the "Damascus Only" box. The agent was napping the next afternoon when the phone jarred him awake. It was the sheikh, highly excited; he barely said hello before announcing that "the Jews have finally gotten Rademacher." Cohen was delighted to hear the news, but it later proved incomplete: the letter bomb had inflicted only minor injuries. The Nazi bureaucrat had escaped.

Mio's field of action was less menacing than Damascus, which was electric with fatal intrigues, but he'd have no sheikh to bring him to Cukurs' home and no one to vouch for him; his target was far more skittish than the approachable Rademacher. And if he found himself in trouble, there would be no radio transmitter with which to summon help.

The plane touched down in Rio. Worn-out from the flight, Mio took a taxi to the Hotel California, checked in, and went up to the hotel restaurant, where he gazed out on the bronzed women walking up and down Copacabana beach and the kites flown by young boys — eagles, butterflies, dragons — snapping and drifting across the cerulean sky. Everyone on the other side of the glass seemed to be enjoying wonderfully carefree lives. Mio gazed at them a moment, nibbling on his fruit plate.

He sent two telegrams, one to Paris and the other to Rotterdam, both with the same message: ARRIVED SAFELY. LOVE TO EVERYONE. That night, he took a "Rio at Night" tour — it was something that Anton Kuenzle might do before — drank *cafezinho,* the sweet Brazilian coffee, and marveled at the profusion of toy stores. The guide explained that Brazilians adored children. "Even the poorest families will insist on having a magnificent christening," she said, "even if they have to mortgage their future for it." Mio posed for a photo with the other tourists on top of Mount Corcovado, and listened to the guide list the amount of concrete (seven hundred tons) that had gone into the construction of the Christ the Redeemer statue atop it. Then it was on to the favelas and a macumba ritual, an African black magic ceremony.

The next day, while on another tour, Mio spotted a sign for the Santapaula Melhoramentos company; he'd come across the name while researching the mission in Paris. The firm had built a yacht club in Interlagos, near São Paulo; Cukurs had his business in the marina. It was a stroke of luck. He walked into the office and found the manager, introducing himself as a businessman exploring investment opportunities in the tourist sector. The manager wasn't in the least suspicious and plied Mio with brochures of the company's holdings. He recommended that Mio seek out his co-workers when he made it to São Paulo. "Tell them I sent you," the man said. Mio wrote down the names and promised to follow up.

He flew to São Paulo, took room 2102 at the Othon Palace, dropped his luggage, then headed out to establish his bona fides as a rich businessman. He rented an orange Volkswagen Beetle and drove to the São Paulo office of Credit Suisse, where he discovered that the Zurich branch had sent along his letters of introduction as promised. He then drove to the headquarters of Santapaula Melhoramentos, where he met with the public relations manager, who had a letter of introduction typed up that would get him into the Interlagos marina. In a matter of a few hours, Mio had left behind a trail of business cards that the aviator, should he grow suspicious, could easily follow.

The next morning, Mio drove to the marina. It was time to take a look at Herbert Cukurs.

At the marina office, he presented his letter of introduction to the engineer who oversaw the company's operations and asked a thousand and one questions. Mio didn't care about the responses; he only wanted to make an impression as a typical Austrian perfectionist who was seriously interested in spending money. As he walked along with the Brazilian, the agent let his eyes wander. There, in the next dock, was a line of pedal boats tied together. It had to be Cukurs' little business. He didn't see anyone who resembled the Latvian, but after a moment he spotted a restaurant with a thatched roof just behind the boats.

Strolling along the dock, he made his way to the restaurant entrance. He walked in, bright sun giving way to a slightly cooler shade, and took a table by a window. The waitress came over, and Mio ordered a cold drink. Through the windows, he could see the pedal boats sloshing together and a seaplane bobbing lazily on its twin floats every time a motorboat raced by and sent a small wake across the water. There was a man bent over one of the boats doing some repairs, far too young to be the aviator. Perhaps one of Cukurs' sons?

Mio sipped his drink. What next? The dailiness of the scene before him, the drowsy heat pressing through the thatch, contrasted with the rising excitement he felt. Should he go out on the dock and perhaps run into Cukurs, strike up a conversation? It was lucky for him that the Latvian ran a rental business; he could always say he was a customer looking to hire the seaplane for a flight over the harbor. It fit his cover. What investor doesn't want to see the full expanse of what he's committing his money to?

Mio was tempted, but he had to remember the file on Cukurs. The man saw Jewish agents everywhere. Better to sink into the boredom of the little dock, let the day run its course, then return.

He paid for his drink and got back in the scorching-hot Beetle.

FOURTEEN

First Contact

MIO RETURNED TO THE MARINA the next day, a Saturday. The weather was beautiful, with just enough breeze to cool the golden-brown skin of the *Paulistanos* who'd flocked to the docks in their bright swimming trunks and tiny bikinis and were now chattering in the restaurant at high volume. Mio ordered steamed oysters in marinade and looked around casually to see if Cukurs had made an appearance. Ahead of him, Cukurs' little boat and airplane concern was doing an excellent business; he could hear the mechanical register, which sat in a little hut on the dock, chime with sale after sale. Mio spotted a young girl sitting behind the register. She was twentysomething, auburn-haired, thin and pretty. Could it be Antinea, Cukurs' daughter, or one of the son's wives? Or just an employee? Mio had studied photographs of the Latvian but none of his family.

Mio walked up to the young woman and said hello in English. She turned and looked him over with her light gray eyes. She recognized his accent. "You can speak German to me," she said. Mio told her he was a businessman scouting for tourist firms to invest in, then launched into a series of questions: "How many customers did they have per day? How did she come to speak German? How many boats did they own?" The questions were designed not only to establish his bona fides as an Austrian entrepreneur but also to bore a pretty young woman who clearly had little interest in the operation of the obscure pedal boat company she found herself working at. She told Mio she spoke German because she'd grown up in Dresden, the picturesque city on the Elbe River. But on the other questions, she had nothing much to say. Instead, she pointed down the dock. "Do you see that tall man, with the white hair?" Mio turned and spotted a figure securing the bobbing seaplane to a pole bolted to the dock. The broad brow, the Roman emperor's nose, the heavy-framed glasses. It was Herbert Cukurs.

The young woman said that the man knew far more than she did about the business. "Talk to him," she said. "I'm sure he'll be helpful." As she spoke, Mio saw her glancing at the pen he'd stuck in his shirt pocket, a ballpoint, still quite rare in Brazil. He plucked it out of his pocket. "I see you like my pen," he said. "It is a Pelikan, made in Germany. Allow me to give it to you, as a memento from our homeland." She was delighted. Mio handed over the pen with a smile, knowing that the woman — Cukurs' daughter-in-law, as he later learned — would show it to her family and talk about the kind businessman who could afford to give away German pens to strangers.

Mio walked toward the older man, who was dressed in a flight suit like the one he'd worn on his trip to The Gambia. Tan, fit, and broad-chested, Cukurs looked far younger than sixty-four. His hair was pure white, brushed back from his brow, which was lined with thick wrinkles. He wore heavy black eyeglasses, and his thin lips were set in a straight line. The face evoked the feeling of a dominant, powerful personality.

"Guten tag," Mio called out.

Cukurs studied the man striding down the dock, taking him in "from head to toe." The aviator appeared calm. *"Auch dir ein guten Tag,"* he said.

Mio came up to him and repeated the same pitch he'd given to the daughter-in-law. "I'm interested in the tourist trade here, and she told me you're an expert."

Cukurs gave a quick answer and went silent. He clearly wasn't interested in tutoring some foreign businessman on Brazil's vacation industry when he had customers waiting. The man was anxious to get back to work; Mio sensed the moment slipping away.

He glanced at the seaplane, rocking on the water. "I hear you fly tourists over the city. Perhaps you could take me for a flight over São Paulo?"

Cukurs perked up. He would, but there was a problem. The flights were priced per person; Mio would have to wait for two more tourists to fill the last two seats before they could go up. Mio waved this away; he would pay for the full tour himself.

"My name is Cukurs," the Latvian said. "Please board the plane."

The aircraft was tiny, a "metal tub with wings." Once he was seated in the ratty leather seat, the hot sun beating down through the windshield, Mio felt just how small the compartment was. There was a profusion of indecipherable dials in a battered metal panel spread out in front of him, and levers jutting up from the floor to his left. Cukurs climbed in after Mio and sat breathing while he prepared the plane, only a foot away, the faded flight suit taut against his thigh. Here was the Jew-killer, close enough to touch. Mio realized he'd hastily stepped into the place where the aviator was most familiar and proficient; he sensed the Latvian's brawny presence next to him. The cockpit began to feel a bit claustrophobic. "With hindsight," he said later, "my hasty decision to get into this airplane bordered on complacent negligence and total disregard of any security measures."

The engine sputtered to life in a puff of blue smoke, which stank of burning fuel. The propeller spun lazily, then whipped into a blur. The noise was earsplitting; Cukurs could barely make himself heard over the roar. With his meaty hand on the wheel, he slowly maneuvered the plane out onto the lake, watching for water-skiers and pleasure boats. The expanse of choppy blue water in front of them was clear, and he pushed the thrust lever forward, the sound of the engine keying higher. The floats rocked on the small waves, and water sprayed across the windshield and knocked against the fuselage as the aircraft gained speed. Finally, the seaplane lifted into the sky.

As he feigned interest in the landscape below, Mio watched his quarry. He was struck by how relaxed Cukurs was at the controls. "One did not have to be an aviation expert to recognize that the man had great technical skills and lived, loved and breathed airplanes," he observed. The agent was searching for a way into Cukurs' life. Every operative has to first answer the question, What does the target want? As the plane's fuselage vibrated in the wind, Mio looked around the cockpit, taking it in. The seaplane was old but well maintained, Cukurs' flight suit was speckled with oil stains, and his glasses, Mio noticed, were carefully taped at one of the joints.

The view of the São Paulo skyline through the windshield was partly hidden behind a hazy line of smog. Cukurs, who obviously knew the city well, began pointing through the side windows to various landmarks, but his voice was lost in the deafening engine noise. Mio nodded and studied the points of interest, sunk in his own thoughts. After twenty minutes, Cukurs banked away from the skyline and dove down toward the lake. Mio grew nervous at the steep angle of descent, but at the last moment Cukurs pulled the nose up and executed a flawless landing, the floats skipping over the surface before sinking deeper as the plane steadied and slowed. When they reached the dock, Cukurs slid back his window and threw a rope to a young man who bore a strong resemblance to the pilot.

Once they were out of the cockpit and standing on the dock, Mio was effusive in his praise. "That was an extraordinary experience," he said. "I thoroughly enjoyed myself." Their transaction was over, and Cukurs had other customers waiting. Mio had to slowly pull the Latvian into his business concerns or begin the first tentative steps toward friendship. If he walked away now, it would be difficult to come back without tripping Cukurs' sensitive antennae.

He let loose with a string of questions: Was his business growing? Did Cukurs expect the flow of tourists to increase in the coming years? Cukurs listened. "Could I invite you to my boat?" he said at last. "I shall be only too glad to answer all of your questions in detail over a glass of fine brandy." First his seaplane and now his boat; it would be the second time Mio confined himself in a small space with the target. But he immediately agreed.

They walked over to one of the excursion boats tied up to the stanchions. Cukurs stepped onto the deck and led Mio into the cabin, where he searched around before finding a bottle of a local brandy. He brought out two glasses and poured them each a generous portion.

"Prosit!" the Latvian said.

Mio echoed the toast, and they drank.

"So what is it you want to know about tourism in Interlagos?" Cukurs said.

Mio launched into his spiel about Austrian partners, expansion to South America, the boom in air travel. But as he went on, he noticed that Cukurs was studying him with "a piercing look."

He finished. Cukurs gave a brief reply, then continued staring at the agent with his blue-gray eyes. "I'm accused of being a war criminal," Cukurs said suddenly.

Mio said nothing; the admission had startled him, but his expression was almost bored.

"Me, a war criminal?" Cukurs went on finally. "I? After I saved the life of a Jewish girl and protected her during the whole war?"

Mio sensed a stratagem. Cukurs was intimating that he'd resisted the Germans, that he was a liberal falsely accused of being a Fascist. If Mio leapt at the opening and said that he, too, was on the side of the Jews, he would show that he couldn't be trusted. "Undoubtedly, he had planned these two last sentences well, in an attempt to place me on the political map, right or left," Mio recalled. "To this day, I cannot comprehend what made him think this was a good tactic." The thought flitted through Mio's mind that perhaps Cukurs assumed he was a war criminal himself, newly arrived in South America, and that he'd come to Cukurs for help.

The aviator was fishing for something, but Mio didn't know what. He fell back on his introvert nature. He said nothing.

Cukurs switched his approach. "Did you serve in the war?" he said. "I'm sure you did." He left the last part of the question—*on which side?*—unasked. The Austrians had been ardent supporters of national socialism; it was probably one of the reasons Yariv had decided that Anton Kuenzle should be a native of that country.

"Yes," Mio said.

"Where, if I may ask?"

"On the Russian front."

Suddenly, "without thinking," Mio found himself reaching up for the top buttons on his shirt, which he began to unbutton. When a few were undone, he pulled the shirt open and revealed a thick, pale scar across his chest. The Latvian stared and said nothing, but Mio could see his mind working. "The scar made a great impression on Cukurs. He must have thought it was caused by a Bolshevik bullet that had penetrated my body on the eastern front when I, with my own hands, tried to block the spread of the loathed communism that had been responsible for his having to flee his own homeland." In reality, the scar was the result of surgery for an abscess that had taken place at a hospital in Petah Tikva, Israel, one of the cities that Cukurs had visited and photographed on his trip to Palestine. Mio hadn't thought about using it beforehand as a visual prop for his war story; the motion, as with an actor deep in his performance, had come to him unbidden.

The evidence of battle seemed to soothe Cukurs. A tentative bond had been established. The two men had served on the same side in the war, and they'd surely both seen—perhaps even done—terrible things. What soldier hadn't?

"And what rank did you hold in the army, if I may ask?"

In the file in Paris, it stated that Cukurs had held the rank of captain in the Latvian Air Force, and the same with the Arājs Commando. For a split second, Mio considered whether to say "major" or "lieutenant colonel," which would mean he'd outranked Cukurs, or go with a lower rank, in order to "foster the mystery that surrounded my character." He decided on the latter.

"Lieutenant," he said.

Mio saw that Cukurs didn't believe him. Anton Kuenzle was clearly wealthy, university educated, and stern; in his silence, he'd enforced a kind of distance and reserve with Cukurs that marked him as a member of the Austrian upper class. Such a man would have gained a higher position than lieutenant, which sat below captain in the Wehrmacht ranks. Mio had taken a chance that this subtle concealment would convince Cukurs that he had something to hide, some unpleasantness with Jews perhaps. A major who'd misbehaved could be on an Israeli list or two, just like Cukurs. And besides, a member of the Austrian upper class would find it inexcusable to brag about his accomplishments to a Latvian who'd served as a captain.

Cukurs seemed satisfied. He asked no more questions, and the talk turned back to tourism. The Latvian's attitude, his entire affect, had changed. Instead of gruff answers, Cukurs spoke at length about the trends he saw developing in Brazilian tourism, which ventures would profit and which would not. The answers chimed with the research Mio had done on the São Paulo economy; Cukurs had a good mind for business.

The two sat in the warm cabin for an hour. Mio could hear the voices from the dock, potential customers for Cukurs' boats, and he saw the Latvian growing restless. "I don't want to disturb you anymore," the agent said. "I can see you have a lot of work to do." Cukurs nodded. "But Herr Kuenzle, it would be really nice if we could continue to discuss our common interests after work." He told the businessman that he lived on the opposite shore of the lake; he invited Mio to his home.

Mio was tempted by the offer, but he had long subscribed to the old Arab proverb, attributed to the Prophet Muhammad: "Patience comes from Allah, haste is from the devil." In Anton Kuenzle's world, someone like Cukurs would be a social inferior. He couldn't appear too eager to meet with a bit player in his firm's global plans. *Let him think of money till he cannot sleep,* he thought. *Let him think of gold and other shiny things.*

"I'm afraid this week is out of the question," he said, rather coldly. "I must go to Brasília and Bahia . . . I think I'll return to São Paulo next week sometime. If I have time, I'll get in touch with you."

"I hope we'll meet again," Cukurs said. "I still have so many things to tell you." Cukurs gave him his home address in the city.

The two men shook hands, climbed out onto the deck, and stepped up to the dock. Mio said his goodbyes and returned to his hotel room, where he wrote a friendly letter to his family. When he was done, he inscribed his report to Yariv in invisible ink. "Made contact with the Late One," he wrote. "[He] swallowed the bait . . . I carry on. Anton."

Herbert Cukurs in the mid-1920s

Cukurs standing in front of the self-built plane he flew to Tokyo, 1937

Riga citizens welcome occupying German soldiers, June 1940.

Zelma Shepshelovich as a young woman

A young Abram Shapiro with his parents

Zelma's rescuer, Jānis Vabulis, also known as "Nank"

A policeman walks above the bodies of recently massacred Jews in a pit near Liepāja, December 1941.

A sign in German and Latvian forbidding unauthorized entry into the Riga ghetto

Tuviah Friedman while he was working as a Nazi-hunter

Simon Wiesenthal

Mio during his time with Mossad

Yosef Yariv in later years

Eliezer Sudit and his wife

Adolf Arndt speaking before the Bundestag about the statute of limitations issue, March 1965

FIFTEEN

The Campaign

THAT FALL, TUVIAH FRIEDMAN INTENSIFIED his fight against the statute of limitations, giving interview after interview and buttonholing Israeli politicians wherever he found them. Civil rights leaders, Jewish organizations, and governments around the world took notice of his warnings that an amnesty for Nazi murderers was approaching the following May. They reacted with shock. The other leading—and far more well-known—Nazi-hunter, Simon Wiesenthal, followed Friedman's path to Bonn to meet with Justice Minister Ewald Bucher. When the two men sat down, Wiesenthal, dressed as usual in an elegant three-piece suit, asked Bucher what exactly was going to happen after May 8, should the amnesty go through. Ducking the question, Bucher began talking of the billions of deutsche marks Germany had spent on Holocaust victims. "This process," Bucher said, "is not subject to a time limit." Wiesenthal tensed. Was it because he was Jewish that Bucher was talking about money? As the German went on, Wiesenthal interrupted. "Herr Minister, the murderer of my mother and the murderers of many of my relations and friends have not been found yet. I don't even know their names. I am addressing the Justice Minister, not the Finance Minister." What would happen on May 8, he asked again. "That's not for me to decide," Bucher said. "That's a matter for the federal government or for Parliament."

Bucher had begun walking back the "secret" opposition to the statute that his assistant had revealed to Friedman. Now he was staking out a more neutral position. The minister ended the meeting without giving Wiesenthal a clear answer as to what the government intended to do as the vote in the Bundestag approached. Bucher had now met with the world's two leading Nazi-hunters, and he'd managed to stall them both. "I realized," said Wiesenthal, "that a fight would be necessary."

Back in Vienna, Wiesenthal decided on a global campaign; it would soon become the largest operation he'd ever worked on. He composed a statement calling the statute "an unprecedented injustice toward the millions of victims of Nazi brutality." Once the murderers knew they would never be prosecuted, Wiesenthal wrote, they would link arms with other enemies of liberty and "spread their propaganda and poison" to the young people of Germany and then abroad. Wiesenthal believed the issue extended beyond the Holocaust. If the West Germans let the killers of six million remain free, the world would be a far more dangerous place.

Wiesenthal's work had earned him enemies; over the years, he'd accumulated enough hate mail to fill an entire room in his suite of offices. One time, the postman delivered a letter addressed only to "The Jew Pig, Austria." Wiesenthal had long ago accepted the fact that he was despised by a large contingent of Austrians, many of whom longed for another Anschluss, but this was a bit much, wasn't it? Was he the *only* "Jew Pig" in all of Austria? Wiesenthal rang up the head of the post office and inquired how exactly they'd determined this particular letter was meant for him. The official replied, in effect, *Herr Wiesenthal, who else could it be?* Wiesenthal opened the letter. The official had been correct.

As with Friedman, his loved ones had suffered because of his high-profile campaigns. Once, when he returned from Milan, where he'd been tracing the path of one of Eichmann's assistants, his secretary told him that his wife had received a threatening call the night before. If Wiesenthal didn't stop his search for former Nazis, the voice had informed her, their daughter would be found and killed.

The work was growing increasingly difficult. Even after the success of the Eichmann case, no prosecutors would take his files, and eyewitnesses "had scattered to the four corners of the earth." The numbers were indeed grim: After the Nuremberg trials, West Germany had investigated about 250,000 individuals for war crimes. Prosecutors had filed indictments for 140,000 individuals, but only 6,656 were convicted of Nazi crimes. Of those found guilty, the vast majority had received minor prison sentences, which were often shortened even further due to good behavior, pardons, or commutations. But even those numbers were misleading. Only 164 individuals had been convicted of the most serious charge, murder. Of those 164, only a few dozen had received the death penalty, and most of them were never executed. Instead, their sentences were commuted to life in prison, and even those men were sometimes released after a number of years as a result of pressure from family or elected officials. The men slipped back into civilian life, their commutations rarely noticed or commented on. Wiesenthal liked to joke that "the dumbest Nazis were those who committed suicide after the fall of the Third Reich." It turned out they'd taken the threat of punishment much too seriously; if they'd only waited awhile, they could have led normal, happy lives.

This was the climate in which Friedman and Wiesenthal conducted their work. The two Nazi-hunters shared a highly charged, often bitter relationship. Friedman was shocked by Wiesenthal's fondness for self-promotion and his slippery hold on the truth. When Wiesenthal concocted an outrageous story about his intimate role in the Eichmann operation — in fact, he'd had nothing to do with it — Friedman became apoplectic and confronted his rival. But the two men were united by the spiritual necessity of their mission, which neither could abandon, despite the toll it took on their lives.

Wiesenthal, who possessed a formidable mailing list built up over two decades, sent his statement to every name on it. Then he waited. The responses came in a trickle at first, and then a wave. The quantum physics pioneer and Nobel Prize winner Werner Heisenberg, author of the famous uncertainty principle, signed on, as did the American playwright Arthur Miller. Professor Joseph Alois Ratzinger, who would go on to become Pope Benedict XVI, sent a letter condemning the German government. Ratzinger's name carried extra weight; he'd been a soldier in the German army while his father strongly opposed the Nazis, and his cousin, who had Down's syndrome, had been murdered by them. Wiesenthal was very pleased with Ratzinger's note. But the biggest catch came in Washington. Robert F. Kennedy, who had just stepped down as attorney general and was perhaps the most famous politician in the world at the time, sent a five-word telegram in response to Wiesenthal's letter: MORAL DUTIES HAVE NO TERM.

It was RFK distilled. Wiesenthal liked the message so much that he adopted it as the campaign's slogan.

Mio flew to Brasília in the heart of the country; it wouldn't do to hide out in São Paulo for a week or to return to fun-loving Rio. Anton Kuenzle had said he was going to Brasília, so he went.

After checking into the Hotel Nacional, Mio was waiting outside for the tour bus that would take him around the capital when a feeling of horror crept over him. Ten feet away, the Israeli ambassador to Brazil, Joseph Nachmias, was walking in his direction. Mio had met Nachmias when he was serving as the police inspector general, and the two had become acquaintances. "The nightmare for anybody working undercover with an assumed identity is to run into someone from their real life," he said. He imagined the ambassador waving at him and shouting, "'What a surprise. How are things? What are you doing here? On your own or is the wife with you?' And all of this in Hebrew!"

Mio ducked behind one of the columns that stood at the entrance to the hotel and watched as Nachmias went inside. What were the chances that, in the remote capital of Brazil, six thousand miles from Tel Aviv, he'd run into an old pal? When the tour bus pulled up, he dashed onto it.

Mio spent the morning marveling at the buildings that housed the Supreme Federal Court, the National Congress, and the president. He was a bit of an architecture geek, and Oscar Niemeyer's modernist structures took his mind off Cukurs. While he was listening to the tour guide describe the construction of the parliament building — an enclosed, windowless chamber whose interior reminded Mio of a submarine — he glanced over and saw, to his amazement, two officials from El Al airline, Israel's official carrier, standing a few yards away. He knew the men well. "This was too much," he said later. He hurried away, averting his face.

On the flight back to São Paulo a week later, Mio had an amusing encounter with the German attaché to Brazil, who happened to be seated next to him and who told him about a Bonn-sponsored project in which Israeli agriculturalists were teaching Brazilians who lived in the immense jungle how to live and farm collectively. "Something like the *kibbutz,*" the man said. "Have you ever heard about *kibbutzim*? Do you know what they are?"

"I have heard something about them," Mio answered, before snapping up the attaché's business card. He'd had two close brushes with people who could blow his cover on this short trip. He reminded himself to be more cautious, returning his focus to the man waiting for him in São Paulo.

SIXTEEN

"Our Own Thomas Edison"

BACK IN PARIS, ANOTHER AGENT, handpicked by Yariv, joined the mission. Eliezer Sudit, who'd spent his early days in the Irgun, a right-wing underground paramilitary organization in Palestine, was thin and physically unremarkable. "If you passed him on the street," said his son Zeev Sharon, "you wouldn't be impressed." But Sudit had a reputation as a fearless, hotheaded fighter. He'd come of age during the struggle for a Jewish homeland. He met his wife, a nurse, when she treated him for several broken fingers, the result of a torture session administered by a rival paramilitary organization, the Haganah. Their love affair played out against the skirmishes and bombings of the War of Independence. "They'd seen the birth of the nation together," said Zeev. "They'd fought for it. They were always telling stories about the early days of Israel." Sudit was the ultimate light/dark character, the mercurial, soulful sabra with the sad past. His wife was the tough one. When he received an invitation to join Mossad, Sudit asked her what he should do. She told him: *If you don't go, I will, and you can stay home and raise the kids.* "He was soft, kind, funny," Zeev said, and had a jovial, lighthearted streak that was rare among Mossad agents.

The choice of Sudit was a calculated risk. He'd been a member of the team that, in March 1952, had constructed a parcel bomb and sent it to the German chancellor, Konrad Adenauer. The plot had been devised by the former Irgun leader Menachem Begin, who was virulently opposed to the reparations Germany was paying Israel, which he considered to be blood money. The bomb exploded before reaching Adenauer, killing a munitions expert instead. Sudit was tracked to Paris and arrested, but later released. If the Cukurs mission fell apart and its agents were identified, recruiting someone who'd once tried to murder the head of the German government wouldn't play well in Berlin. But Yariv felt that he could trust Sudit. He took the chance.

Mio returned to São Paulo and checked back into his hotel. The Paris office had been busy; letters and telegrams from Yariv and the team were waiting for him at the front desk. "Requesting your authorization to close the US$100,000 deal we discussed prior to your departure" read one. Together, the communications painted a picture of a well-funded company doing deals all over the world, with Anton Kuenzle making the final decisions. He composed his own letters in return and went out to mail them.

He then returned to his hotel room and sat on the bed. He still had no master plan for how to maneuver "the Late One" out of the country. There was no way to quickly check in with Mossad; long-distance calls were expensive and difficult, and it wasn't as if there was an operative sitting at a desk in Tel Aviv ready to dispatch a plane or a commando team to São Paulo. He was on his own. How could he get Cukurs out of Brazil without arousing suspicion?

Every good double agent is an amateur psychiatrist. Mio thought about what he'd seen at the marina. He sensed, first of all, that the Latvian was unhappy with his present circumstances. He was the owner and operator of a small, shabby service business where he had to carefully parcel out his time among engine repair, accounting, and bowing and scraping to his customers. Nobody paid attention to Herbert Cukurs; for the tourists who saw him on the dock, he was an aging grease monkey whom they had to speak to in order to have fun on the water. He must miss the days, Mio thought, when the Germans paid for his vodka and cigarettes and when he could do as he liked in the Riga ghetto, where women and jewels and richly appointed furniture were there for the taking. "Then he had been almost an almighty God, at least where the life and death of the Jews of Riga was concerned," Mio said. "He most probably recalled those days fondly when he was at the zenith of his power and glory, treated with great reverence and fear by anyone who saw him riding on horseback through the streets of the ghetto, clasping a heavy Nagant handgun, clad in his black leather pilot's coat." The humiliations of his present life were no doubt painful to him. "Oh, how he must long for those days, which seem now, looking through the window of the shabby ticket office of his run-down boat rental business, like a mirage produced by memory cells with an overdeveloped imagination."

Now Anton Kuenzle had arrived. What did that mean to the Latvian? Mio tried to see Kuenzle as Cukurs saw him. "I tried to imagine what had gone through Cukurs' mind since our so-called accidental first encounter." What was Cukurs feeling? What was he dreaming of? What incidents in his past were influencing his view of the Austrian businessman? Mio quieted his thoughts and tried to feel his way into the dark landscape of the man's inner life.

"Although nothing had been said," he wrote, "[Cukurs] started to think that perhaps I, Anton Kuenzle, businessman, was the embodiment of the unique opportunity, perhaps the last in his life, to considerably improve his standard of living." He could imagine the kinds of scenarios that were just beginning to unspool in Cukurs' mind, visions of new seaplanes, a full-service repair center perhaps, or a yacht club like the one he'd been forced to abandon when the Brazilian Jews had exposed him. Perhaps even that much-anticipated "school of seaplane pilots unlike anything that had ever existed in Brazil." And then there was the aircraft engine he'd already designed, which needed only money for a new factory and employees to get it off the ground. "After all the long and trying years of living on the verge of poverty, perhaps through me he could succeed in once again becoming a respectable man, a man of means . . . just like that glorious period, those wonderful days in 1941."

It was a remarkably good guess. Unknown to Mio, his target was not only an aviator and a mass murderer; once upon a time, he'd also been a novelist. In 1937, Cukurs had published *Starp zemi un sauli* (Between the earth and the sun), in which Aivars, an irresistibly handsome and courageous flier who bears a strong resemblance to Cukurs, invents a new aircraft that will revolutionize air travel. "He was holding the destiny of the world in his own hands," Cukurs wrote. "His imagination and his relentless work over many late nights had made something special that would bring glory to him and immortality to his nation." In the book, Aivars approaches a potential backer. The meeting does not go well, and afterward he broods:

> Now for the third time, he was sitting in this café, smoking his pipe and trying to swallow his sorrow with black coffee . . . He had invented an amazing airplane in his fantastic imagination. However, as soon as he had to execute all his ideas and fantasies, progress stopped. Aivars knew many wealthy Latvians. He told them about his invention, showed them its bright future and possibilities. The capitalists were excited; "We'd found our own Thomas Edison!" they said. But whenever Aivars mentioned that this project required the investment of tens of thousands of *Lats,* the excitement faded and the investors pleaded they had no funds to invest. "Is my invention going to remain only on paper?" [he thought.] "Will it sleep for years until someone else invents it from scratch?"

Mio hadn't read this book — it was never translated from the original Latvian — and he'd perhaps overemphasized Cukurs' greed and underplayed his megalomaniacal ideas, but after spending only a few hours with the man, he'd somehow intuited his dreams.

Nobody saw the potential in Cukurs. His friends had fallen by the wayside; his siblings were thousands of miles away in Latvia. Mio could guess that his children, mortified by the stories in the press and fearful of his rages, rarely asked him about the war or The Gambia or anything in the past. He was probably quite lonely. He clearly longed to be celebrated, as he had been before the war. Why else had he retained his name, at great danger to himself, except for the fact that its four syllables contained all his past glories, without which he would be just an aging marina attendant? Anton Kuenzle could offer Cukurs the two things he desired: money and respect. "Maybe with the help of Kuenzle and his backers' capital," Mio thought, "he will finally be able to pull himself out of this dismal existence, always on the verge of poverty, and become rich, a respectable member of society, and rid himself of the persecution of the Jewish organizations that have been beleaguering him for years."

This would be the bait: a last chance at the golden ring. But Mio had spent only half a day with Herbert Cukurs. How could he be sure he understood the Latvian fully, this deceiver who'd led thousands of Jews to their deaths? He couldn't. "I knew that despite my successes so far," he said, "the most difficult part lay ahead of me."

SEVENTEEN

The Plantation

MIO LEFT HIS HOTEL, got in the orange Beetle, and drove to the address Cukurs had given him in the boat cabin. When he parked the car and surveyed the structure, he saw a house that stood out from the middle-class suburban homes baking under the Brazilian sun. Cukurs and his family occupied "a compound that had been turned into a small fortress." Barbed wire topped a metal fence around the property, and beyond it squatted a medium-size brick house, painted white. Off to the side was a small shack with a metal roof that might serve as a workshop or a shed for Cukurs' gardening tools; it was surrounded by bare ground with only a tree or two to offer relief from the hot sun. Clearly, the Cukurses spent most of their time inside. Between the house and the fence, a big German shepherd was prowling, "exposing its vicious fangs at the sight of the approaching stranger."

Mio spotted a young man, apparently one of the Latvian's sons, standing near the front door. "Good day," Mio called out, "I'm looking for Herr Cukurs." The man said nothing but turned and entered the house; almost immediately, Cukurs opened the door and strode toward the gate. He was dressed in grimy work clothes and he struggled to recognize his visitor. But when he did, a broad smile spread across his face. "Herr Kuenzle! What a surprise. I'm so glad to see you." Mio responded, calling the Latvian by his first name. Mio was very much in charge.

Cukurs seemed very pleased to see the Austrian. It was almost as if he'd been impatiently waiting for the arrival of his new friend ever since that day on the dock. He tied the German shepherd to its leash and unlocked the metal gate before grasping Mio's hand, then invited the businessman into his house.

Mio entered, looking around carefully, curious to see if any of the riches that Cukurs had plundered from Riga's Jews survived in his furnishings. As he suspected, they did not. There were a few pieces of plain, well-worn furniture, but no crystal, no artwork, no grand piano. The Latvian's wife emerged from the interior of the house, a "thin, graceless" woman with uncombed hair. She greeted Mio, shook his hand, then walked back to the kitchen. Cukurs gestured for Mio to follow him and headed toward the living room. On the way, he stopped at a cabinet and opened one of the drawers; inside was a collection of medals and other awards that Cukurs had received. He pulled them out one at a time and explained their origins. Mio saw in the man's face undisguised happiness and pride, especially when he drew out the Order of Santos Dumont medal, clearly the prize of the collection. There were about fifteen medals in the drawer, and Mio saw that many of them were adorned with black swastikas set on red backgrounds. He froze. "At that moment, I preferred not to think of the activities for which Cukurs had been awarded these decorations."

The Latvian opened another drawer. "Look what we have here!" he said in mock surprise. Mio glanced down and saw "an entire arsenal" concealed inside the cabinet: a Beretta handgun, a snub-nosed 7.65 mm semiautomatic pistol manufactured by Fabrique Nationale in Belgium, an older "Broomhandle" Mauser, and a 5.56 mm semiautomatic rifle. It was a formidable array of weapons, and Cukurs gloated over it. "The Brazilian secret service gave me a license for all these toys," he said. "I want you to know that I know how to take good care of myself."

Why is he showing me these guns? Mio wondered. It wasn't necessary for their work; travel consultants didn't need a small armory to do their job. "Was it a warning?" he thought. "Or was this just another attempt to impress me, Anton Kuenzle, former German army officer who had been wounded in a heroic battle against the Bolsheviks on the Eastern Front?" Perhaps in the two weeks since he'd last seen Cukurs, the man's dreams of future riches weren't the only things that had bloomed. Perhaps his neuroses had, as well. Mio nodded and clucked in admiration at the guns, but in his mind he was thinking, *It won't be easy to eliminate him.*

Cukurs led his new friend on a tour of the house, then brought him out to his work shed, which turned out to be very spacious, containing not only the chassis of various marine engines that Cukurs was tinkering with but also a "very old" Cadillac. The car had clearly once been a magnificent specimen but now sat in pieces around the shed. It was another clue, though Mio didn't recognize it then.

"I'm almost finished working on it," Cukurs said, looking at the car. "Another push, a couple of investments and the old lady's as good as new." Mio stared and tried to hold his tongue. The thing was in shambles, not even close to being roadworthy. Cukurs was clearly a dreamer, but Mio could see that he'd already composed a list of projects to finish once the money started rolling in.

There was a small office in the shed, with a desk and some chairs. The two men sat down, and Cukurs poured them some Brazilian brandy. "I have given some thought to our previous conversation," he said after they toasted, "and I believe I can help you in finding interesting investment opportunities." He pulled out a sheaf of photographs and handed them to Mio; they were aerial pictures of São Paulo and its surroundings, shot from Cukurs' plane. Mio was impressed. Cukurs had taken the initiative; he'd found a way that he could be of use to the partnership and had even spent some of his scarce funds pursuing it.

The two men finished their brandy and took Mio's car for a look at the Interlagos area, with Cukurs describing the neighborhoods as they slowly drove through them. "It was a crushingly boring tour," Mio recalled. The area was still largely undeveloped, mile after mile of scrubland, light industrial sites, and suburban housing. Cukurs, who seemed to be watching Mio, gauging his reactions, suggested they head inland, where he owned two big plots of farmland. "They could be suitable for your investors," he said hopefully. "I'll only be too glad to show them to you."

Mio could sense the Latvian's desperation. "So now he was trying to present himself as a landowner," he thought. It might be a way to bring Cukurs into a deal with his fictional partners, and he quickly agreed to the trip. Cukurs suggested a weekday, when his marina business would be slow. Mio sighed; his schedule was nearly full, he said. But after giving it some thought, he managed to carve out a few hours to see the farms. Cukurs smiled and advised Mio to buy a pair of leather boots, to prevent being bitten by any of the highly venomous snakes that lived in the underbrush there. And he promised to get the Cadillac on the road for their little jaunt.

Mio said goodbye and returned to his hotel. He had time to kill. He drove out to the Instituto Butantan, which was world renowned for its collection of 60,000 poisonous lizards, insects, scorpions, and snakes. From behind glass, Mio watched as the scientists milked the venom of the *Bothrops insularis,* the famed golden lancehead, whose hemotoxins were so potent that they caused intestinal bleeding, kidney failure, and the melting of human flesh. It was a beautiful pit viper with golden-brown skin, found only in one place: Ilha da Queimada Grande, or "Snake Island," 110 acres off the coast of southern Brazil, which boasted one snake per every six square yards. When the demonstration was done, Mio left with the other tourists.

On the day of the trip to the farms, the Cadillac, as Mio had suspected, was still sitting in pieces in the shed. With no other vehicle available, Mio was forced to drive the Beetle with Cukurs in the passenger seat. Even with the wind rushing through the open windows, it was hot inside the Volkswagen, and the two said little as they headed away from the coast.

Mio felt he'd made good progress in his seduction, with one caveat: again and again, Cukurs had maneuvered him into situations where Mio was alone and isolated. It went against his belief in DAPA, the Mossad code of multiple plans for every situation; if something went wrong in the jungle, he would have no way to escape. In the mind of every good operative was a small kernel of paranoia that told him he wasn't the one setting the snare, that in fact the snare was being set for him. "Perhaps this was a sophisticated trap laid for me," Mio thought as the tires of the VW hummed on the asphalt. Cukurs had apparently bought his cover story, but how good an actor was the Latvian? He'd fooled so many Jews into believing he was no threat to them; it would be dangerous to underestimate his powers.

Mio was not one of the lethal sabras who did the killing for the agency; his whole career was based on looking like a middle-aged schlub who might find it hard to wrestle a teenager to the ground. But on the way to Cukurs' house, he'd detoured to a sporting goods store and shopped their selection of pocketknives. He now had one, a substantial model with a good blade, sitting in his pocket. Cukurs, he could see, had brought one of his European pistols, along with the big rifle. The knife was as much a psychological defense as an actual weapon, but Mio felt he couldn't go to the jungle entirely unprotected.

They were driving west, toward the town of Piedade. It may or may not have occurred to Mio — he never spoke about it, but it's hard to imagine he missed it — that the long drive allowed Cukurs more than a trip to the country; it also allowed him ample opportunity to see if they were being followed. There was no way to hide a tail on the miles and miles of lightly trafficked roads; any car that followed them would immediately stand out. If Cukurs was indeed trying to check for surveillance, he'd done well.

Above the chugging sound of the VW's little engine, Cukurs

mentioned that the custodian of the first plantation was a sharpshooter who'd once worked as a security agent for the Brazilian president. Mio only nodded. *Why bring that up?* he thought. It might be innocuous name-dropping, or it might be another warning, like the bit about his being able to "take good care of" himself. These sotto voce notes were beginning to add up.

They pulled into Rancho Corujas around lunchtime. There wasn't much to it. "It looked like a small, rundown plantation," Mio observed. What kind of tourist getaway could be built in this scraggly backwoods? Cukurs sensed his impatience and promised that the other farm, Rancho Esclavados, had more potential. Mio drove off, and three hours later they arrived. It was more picturesque, that was true. Lined up in neat rows were 120,000 healthy banana trees, with their stout brown trunks and broad green leaves. The VW bounced along the rutted road as Mio stared through the windshield, wondering if he was being led into an ambush. He noticed a small hut where the custodian might live and a rusting storehouse set back from the road, but all he could hear was the hypnotic drone of insects that rose and fell in the muggy heat. There seemed to be no one around; no one emerged from the hut, and no neighbors wandered by. If Cukurs wanted to kill him and bury his body, he could hardly have chosen a better place. He found a spot and parked the VW.

"Let's take a short walk," the Latvian said. He gestured toward a trail that disappeared between thick bushes. "It leads to a small river swarming with crocodiles." Mio felt Cukurs' eyes on him; the man was studying his responses closely. He quickly agreed. Cukurs hoisted the bag containing the rifle and led the way down the path, the insects chirring in the bush. Mio followed; he felt the knife in his pocket but thought it would be of little use against a long gun, if it came to that. They walked on, the trail darkening ahead of them, splotches of sunlight dancing on the hard-packed soil as banana fronds shifted in the imperceptible breeze. There was no sound of human activity, no engines, no voices. Mio began to sweat.

After four or five minutes, they came to a small clearing. Cukurs stopped. He put the bag on the ground, unzipped it, and pulled out the rifle. He stood up.

"I'm sure you still remember something of your glorious past on the eastern front," Cukurs said, staring at Mio. It was a curious way of putting it. Was Cukurs mocking him?

"Let's have a shooting contest. See the target hanging there, on that tree?" Mio turned. About fifty yards away, a tin plate had been nailed to a tree trunk.

There could no longer be any question about the reason for the journey. The plantation was in the middle of nowhere and lacked every convenience necessary for tourism: sewage lines, telephone, good roads. They weren't here to look at an investment opportunity. They were here so Cukurs could determine if Anton Kuenzle was really who he said he was. "I had no doubt in my mind that this shooting competition was no spur-of-the-moment idea. It was a well-planned move by Cukurs, intended to put me to the test, and maybe even beyond it."

Mio smiled. "With pleasure," he said.

One has to wonder why Mio had refused backup in his first meeting with Yariv on the Avenue de Versailles. Mossad agents almost never operated alone, except in cases where they were working under deep cover, cases in which having other agents nearby was impossible or even dangerous, as with Eli Cohen in Damascus. "Mossad had never done a mission like this," said one former agent. "It was unique." Mio had worked with teams of agents in many of his past assignments. Why change now? Had he wanted to face the Nazi killer on his own, out of a sense of duty to his murdered parents? Or was it bravura, him alone against the Jew-killer? If so, his decision seemed at this moment to have been unwise. And it deviated wildly from his precise, very Germanic methodology; it was as if he'd thrown away twenty years of spycraft in order to go after Cukurs. "He saw the mission as his pinnacle," his son said, "his summit." But perhaps he wanted Cukurs too badly.

The Latvian dug around in the bag and pulled out ten bullets. He slotted the rounds into the breech and raised the rifle to his shoulder. Mio watched. The crack of the gun cut through the insect buzz. Cukurs clearly knew how to handle a weapon. As Mio watched, the sight of the Latvian in a wide-footed shooting stance slowly pressed itself into Mio's mind and merged there with certain other images. "He shot fast," Mio thought, "with the expertise that comes from experience." Experience, that is, that included killing terrified Jews in the forests of Latvia.

Cukurs went through the rounds quickly, and the two walked toward the tree. Ten bullet holes showed in the target, all of them within a few inches of the center of the plate. Mio called it "a decent result." At fifty yards, it was better than that. They returned to the firing spot, and Cukurs handed Mio the rifle. "Your turn," he said, smiling. Mio took the gun, felt its weight, brought the stock up to his shoulder, and sighted the target. "I could see out of the corner of my eye how he followed my every move, waiting to see whether I was an imposter." Mio's hands didn't shake. He truly felt he was an ex–German military officer who'd received a near-lethal wound on the Eastern Front. He took a breath, held it, and fired the first shot, then nine more in rapid succession. He dipped the barrel, and the two marched out to see the results.

They were better than he could have hoped. His grouping was even tighter than Cukurs'. "Way to go, Herr Anton!" Cukurs clapped him on the shoulder. Mio registered the change from "Kuenzle" to "Anton." Ten years earlier, when the Jewish community had first discovered him living in Brazil, Cukurs had told a Latvian journalist that he was "drowning, with no end in sight." He'd even given the reporter his home address, hoping that some like-minded men would come to his aid. Now he felt the call had been answered. "He was almost completely convinced that I was all right," Mio recalled.

The mistrust that Mio had sensed all day vanished. The two of them bantered and laughed. Mio, so attentive to the Latvian's moods, was gratified. "We were, he was beginning to believe, not just comrades in arms but also in fate."

They headed farther along the path. At one point, Mio bent over in pain; something had pierced his foot. When he pulled off his boot, he saw that one of the nails attaching the sole to the leather had worked itself loose. Sitting on the packed earth, he spotted a small cut on the bottom of his foot.

Cukurs bent over him, studying the boot. He offered Mio the gun. "Knock down the bastard nail's head," he said. Mio nodded and grabbed the rifle by the stock. He looked up at Cukurs, his darkened face outlined sharply against the sunlight. "One bullet in the head," he thought, "at point blank range, another bullet in the heart to ascertain death, and the mission is complete." He could kill Cukurs, drag the body into the bushes, and leave it for the wild animals to eat or to slowly decay in the standing heat. He could walk back to the car, drive back to his hotel, pack his clothes, and be on an Air France flight that night. Cukurs' family wasn't expecting him home until the next day, so no alarm would be raised until Mio was back safely in his Paris apartment, his Austrian passport torn up or burned to ashes.

But eliminating Cukurs wasn't the point. He had to be killed in a particular, almost ceremonial way, where the reasons for his execution were clearly laid out and understood. The whole point was to change world opinion, especially German opinion. Just disappearing him on a plantation in the middle of the Brazilian outback, as if he were the victim of a robbery or a kidnapping gone wrong, would be useless. The statute of limitations would stand a greater chance of being enforced, and all the other Cukurs, large and small, would be made whole.

Mio took the gun with thanks and pounded the nail head down, then the two began tramping deeper into the bush. They came to a half-decrepit wooden bridge and stood on it looking at the water flowing underneath, pooling and eddying here and there. Mio spotted the craggy silhouette of a crocodile lurking in the current. There were indeed man-eaters here.

It was getting late by the time they made it back to the car. They found the caretaker's hut and brought their camping equipment inside. Cukurs lit a paraffin lamp and pulled out the supplies for dinner: some vegetables from cans, salami, cheese, black bread. Mio sat and watched him. Outside, the black sky descended, swallowing up the outlines of leaves and bush visible through the windows. There were no lights in the distance, no rooflines; the horizon sank into the murk. It was a moment for a certain kind of travel-weary alienation; Mio must have wondered what star-crossed destiny had brought him here, to the other side of the world, to sit inside a hut in the middle of a plantation alongside a famous Nazi, watching him make dinner. As the knife cut into the tins of food, the isolation and the nearness of Cukurs seemed to press further into Mio's mind. "I could not help but think of my family, exterminated in the Holocaust by murderers similar to the man who sat less than a meter away from me." He thought about his parents especially, and the postcards, supplied by the Red Cross, they'd sent him. The last one was dated June 1943; on the back, one of his parents had written, "Regards from Aunt Theresa." He had no Aunt Theresa; he understood the note to be a code, informing him his parents were now at Theresienstadt, the concentration camp in the Protectorate of Bohemia and Moravia that served as a "retirement settlement" for the *prominente,* famous or important Jews. His father's service during World War I and the Iron Cross he'd won for his courage were the reasons they were bundled aboard the last transport to the camp, where inmates built coffins and painted German uniforms with white camouflage for use on the Eastern Front. Mio learned that his father had been killed at Theresienstadt in May 1944; his mother had lived a while longer before she was taken to Auschwitz.

Cukurs voice interrupted his reveries. "I suggest we go to sleep." Mio stood up and unrolled his sleeping bag on one of the camp beds that were pushed against one wall. He would sleep with his head just a few inches from Cukurs'. Before he closed his eyes, he spotted the Latvian taking out his handgun and slipping it under his pillow.

Mio, a light sleeper, later awoke to find Cukurs standing in the darkened shack, sliding the gun under the waistband of his pajamas. He then walked outside into the darkness. Mio lay there, wondering what the Latvian could be up to, before searching for his own knife. He found it and gripped it in his right hand. Only when his mind cleared and he heard Cukurs urinating against a tree outside did he relax. For the second time, he thought, *It won't be easy to eliminate him.* He reminded himself to state this emphatically in his next letter to Paris.

EIGHTEEN

Paranoia

Tuviah Friedman and Simon Wiesenthal, along with Jewish groups around the world, worked furiously that fall, and their movement began to gain traction. A number of American companies announced a boycott of German goods, prompting Bonn's ambassador to speak out against "excessive agitation" in the United States. A letter campaign was organized. "German embassies and consulates around the world have been bombarded with petitions and deputations . . . ," reported the *Los Angeles Times,* "letters and telegrams in the thousands." And the first protestors appeared on the streets. In the raw cold of a New York winter, dozens of well-dressed men and women marched up and down Park Avenue carrying signs reading NO STATUTE OF LIMITATIONS FOR MASS MURDER. A number of them holding a banner that read JEWISH SURVIVORS OF NAZI PERSECUTION stopped briefly for a photographer from the *New York Times* to snap their picture.

Wiesenthal bought advertising space in newspapers all over the world and published his anti-statute statement, along with the signatures of two hundred luminaries who backed the cause. It was a public relations triumph, and as almost always with Wiesenthal, it had the knock-on effect of burnishing his reputation. The activist even secured the private telephone number of Jackie Kennedy, who made phone calls and donated money to the cause.

Friedman kept up a punishing schedule, meeting with Israeli politicians and organizations and encouraging them to speak up about the issue. Coverage of the amnesty spread through the Israeli press, and vigils and protests began to spring up in Tel Aviv and Jerusalem. Earlier, Friedman had made an appointment to meet with the leader of Germany's Social Democratic Party, Erich Ollenhauer, during the politician's trip to Tel Aviv; he hoped to convince Ollenhauer to publicly oppose the amnesty. On his way into the politician's hotel, Friedman was spotted in the lobby by an Israeli secret service agent guarding the politician. The agent, who knew Friedman well, went pale when he saw him walking toward the elevators. "What on earth are you doing here?" he asked Friedman after hurrying over. "Do you want to assassinate the German Socialist leader?" Apparently, Friedman's reputation for aggressiveness toward Germans had followed him from Poland. Friedman laughed. "Don't worry, Michael," he said, "I have not come to kill him, only to speak with him." The agent allowed him up to Ollenhauer's room but hovered just a few feet away during the negotiations, tensely watching Freidman's hands. Friedman found the whole episode darkly amusing. Despite the rather bizarre atmosphere, he'd managed to convince Ollenhauer to oppose the statute, a major coup for the movement.

But as the campaign expanded, Friedman and the other activists ran into a problem. The Shoah was easily summarized in words, numbers, and pictures; if you simply said "six million" or looked at a photo of the emaciated inmates at Auschwitz, you gained a great deal of visceral knowledge about the genocide. But the statute of limitations was a law written on nineteenth-century parchment; it was hard for the average person to grasp its importance in a few seconds. "Because this seemed like a dry legal issue," Wiesenthal wrote, "it was not easy to transform it into a headline-maker." The campaign had a slogan — RFK's "Moral duties have no term" — but no face, no image.

As the protests heated up, public opinion in Germany remained firmly on the side of the amnesty. Supporters of the statute argued vociferously that jailing more ex-Nazis, many of whom held high positions in companies and public offices, would cripple Germany's effort to rebuild its economy and its government. The Federal Ministry of Justice, for instance, was almost identical to Hitler's Reich Ministry of Justice: the faces were mostly the same, only the nameplates on the doors had been changed. The same could be said of many leading corporations and bureaucracies; there were, for instance, an estimated two hundred "blood judges" who'd served under the Third Reich still on the bench. If Friedman and the others truly wanted to jail every Nazi who'd participated in the murder of a Jew, the country might stagnate. As counterintuitive as it sounded, if the world desired a "new" Germany, it had to allow some Nazis to remain free.

As an argument, it was maddeningly circular: if you prosecute Nazis for murder, you'll get *more* Nazis, not fewer. The obvious parallel was World War I. The harsh terms of the Versailles peace treaty had fostered a resentment and feeling of betrayal among Germans, the famous "stabbed in the back" idea that swept Hitler up in its conspirational fervor. Now the pro-statute advocates were making the argument that Nazi trials would destabilize Germany and induce a Weimar-like interregnum. Behind this theory, some observers sensed a warning: if you don't stop, we'll stop feeling remorseful — or stop *acting* remorseful — and reveal our true anger.

Underlying such arguments lay an unpleasant truth, one that Tuviah Friedman had known for years: Hitler had enjoyed far wider and deeper support than anyone in Germany cared to admit. "Approximately ninety percent of the Germans," an examining magistrate for the Auschwitz trials, Heinz Düx, would later say, "as activists and fellow-travelers, translated the murderous ideology of the National Socialist regime into action." Germans preferred the "guilty few" theory of the Holocaust, the idea that Hitler and a few of his underlings had forced citizens to become accomplices in the genocide of the Jews, because it absolved them of their own guilt. Individual Germans almost never admitted participating in the Holocaust; anyone who did so was regarded as a *Nestbeschmutzer*, one who fouls their own nest. If trials were allowed to go on indefinitely and prosecutors were free to track down every officer and soldier who'd murdered a Jew, a deeper and more disturbing portrait of the Holocaust would inevitably emerge.

Back at his hotel after the trip, Mio wrote a long report to Yariv, focusing on what he now saw as the main difficulty of the mission. "I reiterated and stressed the fact that despite the Late One's age of 64, he was still a dangerous man, alert, physically strong and resourceful." He wanted the team to be ready to fight; Cukurs was neurotic, primed for a confrontation. "Twenty years on the run leave an impact on a man," Mio thought. Each time Mio gained Cukurs' trust, there came a moment when the Latvian snatched it back. "I kept reeling in, then releasing the line, only to reel in again, and then let go. I feared I would have to repeat this act many more times." Mio also addressed the issue of how to finish the Late One off. He was considering a number of methods: poison, sniper, ambush. Each had its advantages, and none could be ruled out at this stage.

Mio finished the letter, sent it off, and went to meet with Cukurs again. After the debacle of the farms, which had revealed no property fit to become a luxury resort, Cukurs proposed another trip, this time to Santos, on the Brazilian coast, about forty-five miles from São Paulo. Mio agreed. He said he wanted to check on the rental market there: how much owners were getting for their vacation homes per week, what they asked for as a deposit. What he really wanted was a chance to test the third option. "I wanted to get him used to entering vacant houses without a second thought," Mio said. He would train the Latvian to walk into strange buildings where an ambush might be possible.

They visited several homes, Cukurs always dutifully following Mio over each threshold. After they'd strolled in and out of the last home, Cukurs suggested a detour to Porto Alegre, or "Joyful Harbor," a tourist destination due south. Mio nodded enthusiastically. "What a great idea," he said. "You have no idea how glad I am we met." He went on in this vein, praising Cukurs for his intimate knowledge of the Brazilian market, and of Portuguese, which any entrepreneur would need in order to expand into São Paulo. He volunteered to buy Cukurs' plane ticket to Porto Alegre and pay his way once they arrived. Careful not to offend Cukurs' pride, he phrased his offer carefully, as a business expense that would be repaid tenfold once the Latvian lent his expertise to the firm. Cukurs seemed pleased.

They met in the city a few days later. Mio was careful to book a flight that would put Cukurs at their hotel first, to remove any thought of a trap from his mind. Now that Cukurs was comfortable walking with him into vacant houses, he wanted to get him used to the idea of meeting him in far-flung cities. When Mio arrived, he checked into their hotel, then went looking for Cukurs' room. He knocked. After a few moments, the door was jerked open. Cukurs stood there with a handgun pointed at Mio's belly.

Mio struggled to keep his nerve. He lifted his eyes from the gun.

"So, what is it, Herbert?" Mio said. "Scared of me?"

Cukurs dropped the muzzle of the revolver and tucked it into his pocket. He seemed aggrieved. "If you had a long nose, then I would've had good reason to be scared. One must always be on alert." Even in this sun-kissed resort, the man was expecting a Jewish assassination squad to knock on his door. Mio brushed the remark aside, but he was disturbed. Would his target ever relax? "He meant every word he said. He was serious, deadly serious."

For the next few days, the two of them toured Porto Alegre, building a thorough picture of the tourist economy. Mio paid for everything — gas, meals, tips — while his Latvian partner bought the occasional cup of coffee, to retain some shred of self-respect. Mio found that his Austrian strategy was working almost too well; Cukurs had become eager to please. Whatever Herr Kuenzle wanted, Herbert sought to provide. He was constantly trying to guess what Mio was thinking so that he could anticipate Mio's needs, much to the agent's annoyance. Mio made it clear he didn't want to chitchat; he kept the conversation on the relative prices of beach houses, and Cukurs followed suit.

As they drove through the city one afternoon, the aviator was droning on about tourism. Mio was half listening when he heard Cukurs mention a name: Josef Kramer. Mio had to work to keep his expression neutral; he focused his eyes on the road ahead. But inwardly he was shocked. Josef Kramer was the infamous "Beast of Belsen," the former commandant of the Auschwitz-Birkenau and Bergen-Belsen concentration camps, who had personally selected Jews for death and gassed eighty prisoners whose skeletons were wanted for an anatomical collection at the Reich University of Strasbourg. Kramer, recognizable by his black-eyed, intense gaze and the scar that curved up his left cheek, had been exceptionally brutal to the inmates, sometimes whipping them until their skin peeled off. Like Cukurs, Kramer had denied everything after the war, telling prosecutors that the stories about him were "a product of [the Jews'] imagination." After a trial, he was hanged at Hamelin Prison in December 1945.

Why would Cukurs drop that particular name into their conversation? It was as if the aviator was taunting him in a language he would understand only if he was someone other than Anton Kuenzle—if he was, say, a double agent—a language made up of the names of executed Nazi murderers and specific types of guns, telling him in a silent undertone, *I know why you're here, and it isn't to make money.* Despite himself, Mio was unnerved.

After a long day of researching the market, the pair sat down for a hearty dinner at one of Porto Alegre's best restaurants. The trip had acted as a preview for Cukurs of what he could expect if he partnered with the Austrian: fine meals, good hotels, respect from waiters and barmen. As they spoke, Mio began to talk about expanding their search to nearby countries. The Brazilian market was obviously ready to boom, but what about Chile, Venezuela, and . . . Uruguay? Perhaps there were even better deals there. They really should check out those countries' tourist sectors and enjoy a bit of the high life while they were at it.

Cukurs reminded Mio that there was a problem. If his name was connected with the company, it could become a liability; the accusations of war crimes might bring bad publicity. "The past does not interest me," Mio said. "I am only interested in the future." The fierce lobbying that Brazilian Jews had conducted to keep Cukurs from becoming a citizen, however, did crimp Mio's plan in another way: Cukurs had never been able to obtain Brazilian citizenship, which meant he had no passport. How could he leave the country without one? There was such a thing as an "alien's passport" that noncitizens could use for travel to neighboring countries, but Mio had little idea of how they worked.

"Is it difficult to get one?" he asked Cukurs offhandedly.

"Not at all," the Latvian said. "It's just a question of paying a middleman a modest sum. US$30 or so."

Mio was relieved. He would supply the supporting letters and documents to show that Cukurs was his business partner and needed to travel outside the country; he also insisted on paying for the passport. "As for the gun," he joked to Cukurs, "you have nothing to worry about. I'll get it through the border controls for you." Ever since the banana plantation, firearms had become a private bond between the two. Cukurs laughed. "Most of the time he kept a poker face," Mio said. "But this time there were laughter lines around his eyes, which sparkled with genuine joy."

They celebrated that night at a Japanese club. When Mio walked in, the maître d' hurried over to greet them and made a fuss over the Austrian businessman. The man led the two friends to a good table; waiters materialized and took their drink orders. Cukurs sat across from Mio, reveling in the special treatment. The Latvian's drink — a Cuba libre — was whisked off a tray and set down in front of him. Mio lifted his cocktail and cried, "Prosit!" The Latvian smiled, and they chinked glasses. Mio was pleased. The friendly attention of the Japanese waiters wasn't an accident; he'd come to the club a few days before and dropped a wad of cash to smooth the way.

The two men watched the dancers go through their routine—it was the cancan—Mio observing his partner carefully all the while. The Latvian tilted the Cuba libre to his mouth and took a small, careful sip, then waited a few minutes before taking another. "He behaved like someone afraid of being poisoned," Mio thought. It was true: Cukurs' wife would later say that when they dined out, Cukurs would inspect the cutlery, then ask the waiter to change it. As he did so, he would watch the restaurant staff to see if there was any strange reaction.

Before Mio left Brazil, Cukurs wanted him to come to his home for some strudel, to cement their new partnership. Mio agreed. The meal turned into a "completely surreal" experience for the agent. The Cukurs family crowded around Mio, their Austrian savior, as they all nibbled on delicious homemade Viennese strudel and drank coffee. The daughter-in-law, whom Mio had met at the dock, translated bits of the conversation for the others. Mio felt like an exhibit in a museum: "I . . . became a real attraction for the rest of the family, who must have forgotten when the last time was that a stranger sat in their living room." They seemed to be starved for company. The Cukurs children appeared to be analyzing each word he said with Talmudic intensity. Did he really intend to make their father rich again, they wanted to know. He assured them he did, and soon. "I must make one more short trip to Uruguay, and them I'm flying back to Europe. I'm going to propose to my associates that we employ Herbert's efficient and dedicated services on a permanent basis." The faces before him beamed with pleasure. At the end of the night, Cukurs walked him out to the street, and Mio told him they'd meet next in Montevideo.

To Cukurs' son Gūnars, it was Mio who acted oddly. "He was a fat guy and did not seem to play any sports," he remembered acidly. "He wore a thin mustache and always spoke with a certain timidity, very reserved. It seemed like he was studying the words before pronouncing them." His mother also found the visitor mysterious: "He always talked about business and avoided any conversation that did not relate to it. His eyes were elusive, and he was sparing in his expressions, calm, calculating." Mio appeared to the family almost as a cliché Austrian: cold, unemotional. And somewhat perplexing.

Mio made it to Montevideo first. The Uruguayan capital was less frenetic than Rio, more conservative, and whiter. Vendors carrying thermoses of hot maté in the pockets of canvas vests slung over their shoulders strolled the streets dispensing the strong, caffeine-rich herbal tea. It was drunk, Mio learned, through a silver straw. Palm trees were everywhere, along with Spanish-style cathedrals and public fountains. Mio watched antique Chevrolets and Fords, a Uruguayan obsession, thread through the streets toward the spectacular half-crescent beach. The faces on the street looked more European than those in Rio; most of the residents were descended from Spanish or Italian immigrants. The city felt to Mio like certain neighborhoods in Madrid; it made him homesick for Europe.

By now, he'd been away from Paris for over a month. His children missed him terribly; he hadn't been able to call even once. After hot, intense Brazil, the calm streets of Montevideo reminded him of Paris, and he felt "great longings for my family." The best the agency could do was relay the good wishes in his letters when they arrived at the post office box in Rotterdam. His four-year-old son became so despondent over the absence that Mio's wife asked one of their Parisian friends to phone the apartment. When the call came through, she announced that his father was on the line; the boy took the phone and, "full of joy, began pouring out his tales." But the voice of the man on the other end was subtly off, and Mio's son sensed something was wrong. "I want Daddy!" he began howling. The experiment was not repeated.

When Cukurs arrived at the hotel, he tapped his lapel pocket and said, "The gun is with me." Despite the pistol, he seemed relaxed, dressed in a new suit that probably cost less than what Mio spent on their first dinner. At a casino they visited, Mio won $200 at roulette and immediately took half the chips and pressed them on Cukurs. "This is your share."

Cukurs began to protest, but the Austrian cut him off. "Herbert," he said, "we're partners, after all." Cukurs took the money.

But Cukurs grew increasingly nervous as the trip went on. On their drive to the resort city of Punta del Este, Mio sensed that he was trying to cover his darkening mood with small talk. He smiled and chatted as usual, but Mio detected an undercurrent. What was the problem? The weather was balmy, the beaches eye-smacking. But Cukurs had spotted something in Punta del Este that raised the hair on his arms. He was convinced he was being followed.

"It happened during the lunch he and Kuenzle had at a large restaurant called Suizo Bungalow," Milda Cukurs later said. "Herbert had noticed a group of four young people who looked Jewish — two women and two men. They were sitting a few tables away from Herbert and Kuenzle, spoke in loud voices and made much noise, as is typical of Jews."

Mio was unable to recall the group. It's possible they were Jewish, or they might have just been dark-haired tourists who'd become a bit boisterous. In any case, it was a bad sign. Away from Brazil, Cukurs was far more suspicious than he was at home. "In my view . . . ," Mio said, "Cukurs was suffering from advanced paranoia." Perhaps bringing him to Uruguay had been a mistake.

In reality, the country had very few Jews living there, and few Israeli tourists visited. But Cukurs apparently suspected that his persecutors were all around him. Back in Montevideo, the two had just sat down for a leisurely dinner when Cukurs eyed the waiter. *"Redst du a bissl Yiddish?"* he asked him. (Do you speak a little Yiddish?) The waiter's face expressed only confusion. Cukurs switched back to Spanish.

Their research in Uruguay completed for the moment, Mio announced that he was flying back to Europe, where he would tell his partners about the rich opportunities awaiting them in South America and about their new point man in the region, Herbert Cukurs. "Tourism is about to blossom here," he told Cukurs, "and I want to be in on it." He was going to propose that the company dedicate significant sums of money to the Brazilian/Uruguayan venture. When he returned in the New Year, they'd get started. Cukurs was delighted.

NINETEEN

The Sabras

THE SUMMER TANS OF THE PARISIANS were faded by now, the brittle skeletons of leaves mashed into cold puddles of rainwater on the sidewalks. Winter was around the corner. After his plane touched down, Mio went straight to his family's "mission" apartment and was met at the door by his children, who were overcome with excitement. "Now I'm staying put for a few months," he told them, "together with you."

He played with his four-year-old and his siblings, then went up to the Caesarea flat and knocked on the door. Yariv let him in, and they began going through the large file on "the Late One" — they never spoke the name "Cukurs" in their briefings — that had accumulated over the past weeks. There was a section that listed the minutiae that had to be mastered before the team could fly to Uruguay: the availability of rental cars (which Mio had some trouble with on his original trip), the exact requirements for renting a property there, the papers one needed to book a room at a Montevideo hotel. Once the team arrived in Uruguay, they would be on the clock, with only a short time before the Late One arrived. Mio had brought sheaves of material with him: maps of Montevideo, lists of what was needed at Uruguayan passport control, photographs and descriptions of his encounters with Cukurs, diagrams of distances between hotels, business cards from the people he'd met. It was like sending a librarian out into the field; the man had brought back an archive.

There was one point that Mio made again and again to Yariv and Sudit: "I emphasized repeatedly that we were dealing here with a highly mistrustful man, acutely vigilant, armed with a gun, and of great physical strength." It was difficult for Mio to convey, in this apartment thousands of miles from São Paulo, what it was like to be around Cukurs. He wasn't only phenomenally strong, but he also gave off an almost contagious feeling of suspicion that never wavered. Mio had included the warning in his letters, and by now the others were growing bored with the subject. They'd faced far more imposing targets—lithe young Arabs who'd vowed to kill Jews and would think nothing of blowing themselves to bits if they could take a Mossad agent with them. He could sense the skepticism in their eyes.

They went over the possible ways of executing the Late One. Mio held the floor. He discounted the sniper idea; it would be loud, public, and messy. Cukurs' body would be discovered immediately after the shot. How, then, would the Mossad team get out of the country? It was impossible to say how capable the Uruguayan police were, but they might check the airports, and if there was a description of Mio, it could be unfortunate. Strychnine or some other poison was another no. Mio told Yariv and the others about his dinner with Cukurs at the Japanese restaurant. They wouldn't be able to get enough of the stuff into him at one time to guarantee his death. And they didn't want him waking up in a hospital and describing the mysterious Anton Kuenzle to the police. Finally, Mio gave the team his recommendation. "I believe that the best way is to lure him into a house where an elimination unit would be waiting for him," he said. "From my intense and intimate acquaintance with the Late One, and from the numerous hours we have spent together, which have included many visits to various rental sites, I am convinced that he will enter any house I do. He trusts me."

It took only a moment for Yariv to second the idea. "Mio was there, he knows the man," he said. "We'll lead him into a trap, overpower him, read him the verdict in the name of his 30,000 victims, and carry out the execution there and then." If all went well, Cukurs would die in roughly the same way as the Jews of Riga had died. He would be betrayed by a trusted friend and have a moment to think about his own death before being executed at close range. A sniper would be as impersonal as a gas chamber; this would be close-up, intimate, the way Cukurs and the others had killed their victims in the Bikernieki forest.

Based on Mio's description, they would need at least four men total to bring the Late One down. Yariv contacted a Mossad agent named Moti Kfir, who'd grown up tending sheep on an Israeli farm and later attended the Sorbonne, where he'd studied history. At Mossad, Kfir worked as the director of the school for special operations. He agreed to join the team. The next choice was Zeev Amit, a former paratrooper and devotee of martial arts. "He was a very brave guy, typical sabra," said Eliezer Sudit's son. "He was proud of the country, knew what he wanted, had no self-doubts." Amit had served in Unit 101, a ruthlessly effective, controversial special forces team commanded by future prime minister Ariel Sharon. Its members were handpicked from the rural kibbutzim, men close to the soil, and they specialized in reprisal raids on Arab infiltrators who regularly crossed the border from Jordan and attacked Israeli villages. Critics accused the 101ers of killing Arabs indiscriminately, especially during the 1953 Qibya massacre, in which at least sixty-nine Palestinians died when Sharon's men waded into the heavily guarded village and began clearing houses by tossing grenades and spraying the insides with live rounds. But most Israelis considered them intrepid soldiers who lived far out on the knife's edge. Sudit was delighted when Amit joined the operation. "They were a team of two," said Sudit's son. "They were always making jokes, even when perhaps they shouldn't have."

Secretly — he told no one on the team about it — Sudit had a personal connection to the mission. His mother and grandmother had both been born in Riga, and many of his extended family had flourished there before the war. They had all died during the *aktions*. Perhaps Cukurs had played a part in their fate; perhaps he'd shot an uncle in the back or stood in the doorway as his cousins futilely gathered their knapsacks on November 30. Perhaps not; the details of their deaths didn't make it out of Latvia. After hearing about the target, the agent had arrived at his own conclusion as to why Cukurs needed to die, one rooted in Jewish history, one might even say in Jewish thought. Sudit believed so firmly in the righteousness of Israel that he'd killed for it before. But no, he wasn't doing it for the state, not in the final evaluation. And it wasn't revenge for his dead loved ones in Latvia either. "There was no hatred of Cukurs," his son said emphatically. *"None."* For Sudit, the mission was too important to be fed by a vendetta. As he listened to Yariv describe the final act of the operation, Sudit didn't see the faces of his relatives, as Mio had. Uruguay would be about something else for him.

It was an idea that filled him with something like joy. "If someone killed Jews in the future, other Jews would find them," his son said, summarizing his father's thinking. "No more killing Jews and doing nothing!" This one death would allow other Jews, Jews in distant countries, future Jews, to exist, because for the first time Jews would do what other people had done since the beginning of time: obtain justice for their murdered loved ones, which would slow or even prevent the murders of many other Jews. After two thousand years, there was to be a change. "It was something new!" his son exclaimed. "If you kill Jews because they are Jews, we'll find you." The idea, he said, struck his father and the other men with the force of a revelation. This was how Eliezer Sudit came to see the mission: Israel was risking a great deal and Mossad was undertaking its first execution of a Nazi war criminal in order to give life to the Jewish people. And it pleased him deeply to be part of it.

The team was finalized, but there was one problem with it: God had made most of the Mossad team on the small side. Yariv and Sudit were short; Kfir was of average size; Amit, the former paratrooper, had more muscle mass than the others but was hardly strapping; and pudgy Mio was meant to be an observer only. The four members of the kill team knew how to handle a gun, but they had no idea how to subdue the Late One without using one. Yariv began searching for someone to teach them.

On November 9, a letter addressed to Anton Kuenzle was delivered to the mailbox in Rotterdam.

> Dear Herr Kuenzle,
>
> Following your request, I hereby enclose my flight ticket from São Paulo to Montevideo and back, as well as a few other receipts for various expenses. I hope you arrived well in Rotterdam, and that all is well with you. Hoping to see you soon.
>
> With friendship,
> Yours,
> Herbert Cukurs

The letter was forwarded to Mio in Paris. He was pleased. "So all the work in South America had not come to naught," he thought. "Cukurs had developed a dependency on me." But he didn't rush to respond, putting the letter aside for weeks. Haste was the devil's work.

Soon after the letter arrived, Yariv found his trainer: a naturalized Israeli named Imi Lichtenfeld. Lichtenfeld had grown up in Bratislava, Czechoslovakia, before the war, the son of a hard-nosed police inspector who moonlighted as a jujitsu master. Lichtenfeld's father taught his son gymnastics, boxing, and wrestling; as a young man, Lichtenfeld appeared bare-chested in publicity photos, looking like a lean, sculpted middleweight. He joined the Czechoslovakian national wrestling team and won championships across Europe. Even at five-foot-six and 150 pounds, he was a relentless, punishing opponent.

With the rise of Nazism in Germany, anti-Semitic gangs began swarming into the Jewish quarter in Bratislava. If they caught a Jew alone, they would attack him, leaving him bludgeoned on the pavement or bleeding out from stab wounds. The Fascist government in Bratislava offered terrified Jews little or no defense against these mobs; the perpetrators went unpunished. One afternoon, Lichtenfeld collected the boxers, wrestlers, and amateur bodybuilders he knew from the quarter and marched them out to confront a large group of Czech men who'd arrived at the gate. There were hundreds of young thugs waiting; the baying of their voices thrummed the air like a soccer crowd's. Lichtenfeld, carrying a large blue-and-white flag adorned with the Star of David, led the Jewish athletes out of the neighborhood. When he spotted the Czechs, he began waving it back and forth in front of them like a red handkerchief. "Who is the man to take down my flag?" he shouted. One Czech emerged from the crowd and came toward Lichtenfeld, grabbing at the flagpole. Lichtenfeld snatched the man's arm, hoisted him up, and threw him over a cemetery wall. The rioters ran away.

Later encounters turned into ultraviolent melees. The Czechs brought knives and even revolvers to terrorize the Jews; their confrontations were accompanied by screams, gunshots, and the thud of bodies on stone. A wasted move could mean a smashed collarbone, a severed vein. To survive, Lichtenfeld began developing a street-fighting technique called Krav Maga ("close combat" in Hebrew), which allowed the Bratislava Jews to inflict the most damage in the shortest possible time. His maxims were simple: "Use natural movements and reactions"; "Always attack, even when defending." What he taught was ruthlessly real-world. "Unlike traditional martial arts," one academy proclaimed in an advertisement, "Krav Maga makes no attempt to transform you into a spiritually enlightened warrior."

Lichtenfeld left Czechoslovakia the year after the Nazis invaded, losing an eye in a wretched, two-year journey toward Palestine, which he finally reached in 1942. (His mother died at Auschwitz-Birkenau, and most of his other loved ones were consumed in the

Holocaust.) Word of his expertise in the vicious arts spread. In 1948, the Israel Defense Forces adopted Krav Maga for training its recruits and named the Czech newcomer chief instructor for physical fitness. Lichtenfeld, dressed in a white karate uniform, spent decades teaching generation after generation of young soldiers how to gouge, hit, and maim.

When he was contacted by Yariv, Lichtenfeld agreed to train the team. He was incurious about what the men would do with the skills he taught them. It was all, in the end, the same thing for him: he assumed they were going to hurt those who wanted to hurt Jews. "Imi never asked why," Mio said, "nor did he dwell too much on the request he had received, namely to take a group of men and teach them how to fell a man with one aimed blow."

One can't help but suspect a subconscious motive in choosing Krav Maga. Even Israelis believed that during the war, pale, slight Jewish men had been led to the death pits by strapping, robust killers, something many of the sabras regarded with shame. Now a group of tough Israeli Jews planned to find one of those murderers and strike him to the ground with their bare hands. It would be a tactical way of ensuring silence and stealth, but one has to imagine it was something else as well: a display of Jewish masculine power.

Not everyone, however, was pleased with Lichtenfeld or with the plan to immobilize Cukurs before reading him his death sentence. When the outspoken Sudit heard about it, he was appalled. "It's not a movie," he fumed. "Bring a gun!"

TWENTY

"Certain Categories of Murder"

IMI LICHTENFELD DROVE THE MEN HARD; the sweat flowed freely as they practiced the same maneuvers over and over again. Watching the sessions, Mio thought the instructor moved "with the graceful movements of a dancer," and it was true. His feet slid across the floor like a tango partner as he avoided knife thrusts and clubs dropping toward his head. Now fifty-five, the Krav Maga teacher could appear macabre in his white karate tunic. He had a glass eye, and half his face was frozen into an inert mask from an injury he'd sustained during his perilous journey to Palestine. He spoke in a deep baritone but with a slight lisp caused by the slack left side of his mouth.

The method did require slight adjustments for this mission. Having been invented to save Jewish lives, Krav Maga was defensive in nature. Its core philosophy was "Do whatever is needed to cause as much damage as possible to your attacker and get away safely." But in Uruguay, the Mossad men would be the aggressors. They practiced disarming a target, as Mio had told them Cukurs carried a gun everywhere. In Krav Maga, if there was a gun involved in a fight, one always assumed it would be fired. But the agents wanted to immobilize Cukurs before he could draw it. They practiced a powerful blow from the back they hoped would bring the Latvian down, perhaps even kill him.

Yariv, Kfir, Amit, and Sudit practiced front kick after front kick, pivoting their hips toward the target to increase their power. They paired off and rained blows down toward the heads of their opponents, then switched off and became the defenders. Lichtenfeld believed that the knuckles were too delicate for a good smash to the face; he preferred the meaty base of the hand or, even better, the elbow. Balance was key. Kick, return to position. Kick, return to position. Kick. They drilled and drilled, broke for water and a quick rest, then drilled some more.

What made the experience that much more exhausting for the men was that Mio, who wasn't even doing the knee-thrusts and the rest of it, kept reminding them that they were about to face a raging animal. "Remember, the Late One is a strong man," he told them more than once, "and he's always armed." Even Yariv grew impatient with him. "[Mio] exaggerates a bit," he told the others. "We'll overcome Cukurs without any problem."

It was Mio's turn to be annoyed. The Israelis always believed they could call on their mystical inner resources to solve any problem. Typical sabra bullshit. He was worried they were going to mess up his operation.

After letting Cukurs' latest note sit for over a month, Mio decided it was time to answer it. He composed a letter dated December 31, as if he was finishing up some half-forgotten business before the New Year. He wrote:

> Dear Herbert,
>
> I thank you for your note, and for the receipts you sent . . . Everything here is running as it should, and as far as business is concerned, I have nothing to complain about. My associates are very pleased with the outcome of my trip to South America . . . I intend to return . . . to

implement and realize various investment plans, including the ones you have taken some part in. I intend to take care of some matters in Uruguay and Chile during this trip. I shall therefore be grateful to you if you could manage soon after receiving this letter to make the necessary arrangements for a business trip to these two countries, and obtain the appropriate entry visas.

A happy new year to you and your family. Hope to see you soon.

Your friend,
Anton Kuenzle

He sent it airmail and waited. The reply arrived in the Rotterdam box on January 20.

Herr Kuenzle,

I completed all the arrangements as you asked. I have the passport, including the visas for Uruguay and Chile. Waiting for your arrival, and for further details concerning our joint business trip.

With friendship,
Herbert Cukurs

The Late One was in play.

Between their workouts, the men began taking Mio's voluminous documents and wrestling them into a working paper. Each member of the five-man team would have a specific role: Mio would be the lure; Yariv the commander; Sudit the advance man, who would find a house for the execution; and Amit and Kfir the body men, there to make sure Cukurs was contained and passive as he listened to the words to be read out over him. The working paper included maps, timetables, and the identities the men would assume. Each man was instructed to memorize the basics of his cover — name, country of origin, business — until he could repeat them in his sleep. Each received a new passport, freshly made in Mossad's headquarters. Sudit's cover name was Oswald Heinz Taussig; like Mio, he would be traveling as an Austrian. One of the men visited a currency exchange and bought a large quantity of Uruguayan pesos, while the others studied street maps of Montevideo, drawing routes in the working paper. And they all learned a set of Spanish and Portuguese phrases, just enough to get them around the city and out of trouble in case the mission hit a snag.

Soon after Cukurs' letter arrived, as the men were finalizing their travel plans, they received news from Damascus. Just after 7 a.m. on January 24, 1965, military policemen and undercover agents armed with Samovar machine guns had broken into the apartment of Kamel Amin Thaabet—the pseudonym of the Mossad operative Eli Cohen—and found the agent sitting on his bed with his right index finger on the Morse key of his miniature transmitter, tapping out a message to Tel Aviv. Despite his shock, Cohen managed to toss a bottle of acid on the ciphers and code sheets before one of the Syrian agents placed a gun to the back of his neck. The forty-year-old operative was arrested and tortured, and eventually made a full confession.

Cohen had learned that dozens and dozens of ex-Nazis were living comfortably in Damascus, including Walter Rauff, who had helped develop the mobile gas vans that had asphyxiated thousands of Jews with carbon monoxide. But, besides the slight wounding of one war criminal, he hadn't been able to extract justice from any of the killers. Now it was he who was scheduled to die. Four months later, among the gray minarets of Marjeh Square, Cohen, wearing a white sheet on which the guilty verdict had been written in a neat Arabic hand, was led out from the central police station and hanged in front of 10,000 Damascenes.

The arrest "inevitably influenced our mood," Mio said with some understatement. If the team was caught planning a killing in Buenos Aires, the mission's staging ground, they'd face serious consequences—not hanging, most likely, but long prison terms. Argentina was a place of cross-currents for Jews; its government was the first in South America to allow them to hold public office, but it had also accepted fleeing Nazis, including Adolf Eichmann, into the country. The neo-

Nazi Tacuara Nationalist Movement urged attacks on Jews and was partial to bombing synagogues; after Eichmann's execution in 1962, the MNT (from its Spanish name, Movimiento Nacionalista Tacuara) went on a rampage, committing violent anti-Semitic acts such as kidnapping a Jewish teenager and carving swastikas into her skin. If Mio or one of the others was caught in Argentina, things could turn ugly.

Pressing on, Yariv assembled the team in the Caesarea apartment to iron out the last details. "The day after tomorrow, Mio will wire Cukurs about his return to South America," Yariv told them before turning to Mio. "Where and when would you like to meet him?"

"In the transit lounge of São Paulo airport," Mio said. The plane would connect in the Brazilian city, and meeting in the lounge would mean that Mio could avoid having to get another Brazilian visa.

When all was worked out, the men filled glasses with wine or champagne. Yariv looked at Mio and raised his glass; neither of them was much of a toastmaster. "Best of luck," the team leader said to the four men gathered around the table. The others chimed in. Someone called out *"L'chaim,"* and they drank.

As the team prepared to fly to South America, anti-statute protests spread across the United States and Europe: Atlanta, Boston, Chicago, Cleveland, Detroit, Houston, Kansas City, Los Angeles, New Orleans, San Francisco, Washington, Tel Aviv, and Paris would all see marches that spring. The signs varied, among them WE WILL REMEMBER BUCHENWALD and THERE IS NO ATONEMENT FOR THE MURDERERS OF CHILDREN. Six hundred stood in a silent vigil in London, which was interrupted only by the "subdued sobbing" of the Holocaust victims among them. "I am the only survivor of Bergen-Belsen of my entire family," one woman told a reporter in

Toronto during a march there. "I am so lonely without my relatives." The *New York Times* editorial board threw its weight behind the protests. "Their barbarous acts must not go unpunished," the editors wrote of the ex-Nazis. "This is not a matter of vengeance, but of justice."

What would happen if the statute went into force? Simon Wiesenthal received a photocopy of a letter from a Nazi war criminal, in hiding in South America, to his family. In it, he promised that once the statute was enforced, he would return to them. Wiesenthal went public with the letter. A Toronto synagogue received a warning from one Helmut F. Dieskau, commander of something called "the Union of Fascists (Canada)." When the amnesty passed, Dieskau wrote, it would mean "freeing thousands of anti-Jewish patriots of the threat of persecution and terrorism at the hands of blood-thirsty Jews." After May 8, he continued, neo-Nazis and their sympathizers would begin "a tremendous, unprecedented campaign, including physical attacks against pacifists, Jews and Communists." There were other, more pointed threats. A batch of leaflets, postmarked from a town near Bonn, arrived at German news agencies and the offices of the legislators who would soon be debating the issue; any member of the Bundestag who voted against the statute, the leaflets said, "is hereby sentenced to death." The message was signed "Oberbrigadeführer of Defense Area 3."

The German government, led by the indecisive Chancellor Ludwig Erhard, struggled to find solid ground on the issue. Erhard stated that his personal preference was for continued prosecutions of "certain categories of murder," a compromise that pleased no one. Erhard was beleaguered and increasingly isolated. His cabinet voted "overwhelmingly" to let the statute go into full effect, and his coalition partner, the Free Democratic Party, "dominated and indeed blackmailed" the chancellor on the issue. Erhard faced blowback whichever way he turned; the spreading controversy over the amnesty, visible in the protests that regularly made headlines around the world, was beginning to crimp the country's foreign policy and darken its international image. Despite the billions of deutsche marks it had spent on reparations, the "new" Germany was beginning to look like the old Germany. Still, the latest domestic polls showed that 57 percent of Germans supported the amnesty.

The break did not always follow predictable lines. Adolf Arndt was one of the politicians who backed the statute. The son of a law professor, the sixty-year-old was a curious figure — a tall, "ugly, elderly, schoolmasterish man" with a serious legal mind. Arndt had refused on principle to join the National Socialist German Workers' Party before the war and was subsequently fired from his position as a judge. Instead, he used his skills as a lawyer to help Jews escape to the United States, Britain, and Switzerland. After being categorized as half-Jewish himself, he was interned in a work camp in 1943 and forced to perform compulsory labor; the following year, he was arrested and sent to prison. And yet now he'd emerged as one of the leading voices in support of the statute, arguing, in effect, for a limit to the guilt of those who'd persecuted him and his family. He took this stance despite belonging to the leftist Social Democratic Party, whose members overwhelmingly opposed the amnesty. In fact, he was the only member of the SPD to support it.

Justice Minister Ewald Bucher was still the face of the pro-statute contingent, and early in 1965 he made a startling announcement: if the amnesty was defeated by a vote in the Bundestag, he would step down. BUCHER STAKES JOB ON STATUTE read a headline in the *Jerusalem Post.* Britain's *Guardian* reported that with the maneuver, the blond minister had become the "most vociferous opponent" of the anti-statute forces.

Twenty-six hundred miles away, in Haifa, Tuviah Friedman mulled over the news. In their meeting months before, Bucher's personal assistant had promised that the minister secretly opposed the amnesty on such men and that Bucher offered words of support to the pro-statute side only "in order to calm the former Nazis." Friedman had been hugely gratified to learn that his leading opponent on the issue was, in fact, a secret ally. But now Bucher's true position had been revealed. He'd supported the amnesty all along and was now staking his political career on its enforcement.

It appeared that Friedman, master interrogator and relentless pursuer of Nazis, had been played for a fool.

TWENTY-ONE

The Camera

ON JANUARY 28, MIO FLEW from a winter-shrouded Paris directly to São Paulo. Later, the rest of the team would jet to Buenos Aires, where they would gather before making the trip to Montevideo. But Mio needed to check in with Cukurs, to make sure the hook was firmly set. He waited until all the São Paulo passengers had left the plane, then headed toward the front of the aircraft, a funnel of warm air pushing through the doorway. Stepping out on the top platform of the aircraft stairs, he was startled to hear someone calling his name.

"Hello, Herr Anton! Herr Anton, I am here!" It was Herbert Cukurs, his voice raised up an octave in excitement. Mio looked down in confusion toward the bottom of the steps and saw the Latvian, wearing a dress shirt and brightly colored tie, waving with one hand while the other held a movie camera pointed up at him. Mio froze for a split second. He hadn't been prepared to meet Cukurs just yet; the man had somehow gained access to the secure area. And he had a camera. "What was going on here?" Mio thought. "This meant that Cukurs was actually filming me, immortalizing my image on 8mm celluloid!" A flutter of panic shot through his mind. Why would Cukurs do such a thing? Mio brought his hand up to wave back at the Latvian and held it in front of his face, trying to block him from getting a good shot. But he knew that the moment he'd stepped out on the aircraft stairs, Cukurs had captured his face clearly.

The man was never fooled, never able to fully give himself over to trusting anyone. "He'll never be rid of his suspicions," Mio thought. "I was fully convinced this was another one of Cukurs' security exercises, intended to test my honorable intentions. One false move that might be interpreted as an attempt to hide from the camera and all my months-long efforts to win his trust would come to naught." His hand still held high, the agent walked quickly down the stairs, trying to appear as though he was greeting an old friend while shielding his eyes from the bright sunlight. The surviving footage shows how awkward he appeared; the blocking maneuver looks completely unnatural. Mio had been caught badly off guard.

Mio reached the bottom of the stairs and extended his hand to Cukurs. "I'm so glad to see you, Herbert," he said.

Cukurs let the camera drop to his side. "Me too. We are so lucky. We almost missed one another. I received your telegram only this morning."

They walked out of the heat and into the transit lounge, found seats in the small restaurant, and ordered *cafezinho.* Mio composed himself. He announced that his partners had given him the green light to find and buy investment properties in the region. He was about to launch into a description of the glowing future that lay in store for Cukurs — it was the final sprinkle of salt on the bait — when the Latvian, his eyes fixed on the table between them, interrupted.

"I have a little problem. My papers aren't ready yet."

Mio was speechless. What was Cukurs playing at? The plan was in motion. If the target couldn't make it to Montevideo, it would be difficult to reroute the Mossad members and come up with a new strategy on the fly.

"How could you do this?" he half shouted at Cukurs. "And you say you want to be a businessman? I'm a businessman and I stick by my word!" Cukurs had sworn to him he had all his papers in order and was ready to begin. Clearly that wasn't the case. Mio's deep Germanic hatred of laziness and disorder came through in his voice, which cut through the humid air. Cukurs wilted under it "like a miserable and guilty poodle."

The two sat there. Was Cukurs getting cold feet? Mio wondered. He snapped off another sentence about how disappointed he was with his protégé, then went silent. Cukurs said nothing. His usual bonhomie had drained away.

After a moment, Mio took a deep breath and spoke to the Latvian a little more kindly. Cukurs promised it would never happen again, and Mio accepted this.

Now firmly in control, Mio laid out the future moves to Cukurs. The Austrian partners had decided that Brazil, with its heavy-handed military government apt to impose financial or travel controls at any moment, wasn't a secure place to invest. They were looking elsewhere, to Chile or Uruguay. In the next few days, Mio would fly to those countries, and he needed Cukurs by his side. This was Mio's main point. Cukurs had become the linchpin in the whole operation. The partnership was depending on him to guide them to the right investments.

Cukurs' affect changed. He accepted the compliments like ointment after a whipping; when Mio handed over an envelope stuffed with currency for expenses and plane tickets, he practically beamed. Here was physical proof that he was on the verge of a new life. And yet, as he tucked the money away, he looked at Mio. "Do you by any chance have the names of the hotels where we'll be staying in Montevideo and Santiago?"

Mio frowned. *It never ends,* he thought. "Do you have a pen?"

When Cukurs pulled one out of a pocket, Mio rattled off the list of hotels. He tossed in their addresses and telephone numbers, which he had memorized for just such a scenario as this. Cukurs copied them all down.

An announcement floated through the cafeteria: *"Passengers on Air France flight from Paris to Buenos Aires are requested to go to terminal A . . ."* Mio stood up. "That's it, Herbert. I must go. Let me know when you're arriving, and don't let me down." He clapped the Latvian on the shoulder and headed off to Terminal A.

It had been a short but eventful meeting. Dealing with a neurotic was exhausting, Mio found. He now had to consider the possibility that the Latvian would pause at the entrance to the house in Montevideo and refuse to go in.

As it turned out, Mio was right to be worried. After returning from the airport, Cukurs gave the movie camera to his wife, Milda. "If something happens," he said, "this is my killer." She put the camera, with its undeveloped film, away for her husband's return. Cukurs was increasingly prepared for a confrontation with the shadowy forces he believed were following him. "If someone comes after me," his daughter, Antinea, remembers him telling the family, "I will fight until the death."

In Buenos Aires, complication after complication arose. The South Americans were worse than the French; workers at the post office and the telephone exchange had gone on strike only hours after Mio's arrival. He was left incommunicado, with no way to tell Yariv that the plan was on track. And Cukurs, still in São Paulo, had no way to tell Mio that he'd obtained the correct papers for leaving the country.

To kill time, Mio took a pleasure flight to Mar del Plata, a beach resort on the Atlantic. On the way there, the plane began jerking and sliding as it flew through a powerful thunderstorm. The frame rumbled and shrieked as the aircraft was shoved by strong winds, and lightning branched in the tiny windows. Finally, the pilot turned around and headed back to Buenos Aires. It seemed like a bad omen. Mio took a train to the resort.

Mar del Plata resembled a movie set of a once glamorous resort gone to seed, a place where a film noir hero comes to hide from the syndicate. The umbrellas were tattered, the waiters' white shirts stained, their cuffs dingy. It dampened Mio's mood even further. "My longings for home," he said, "reached new heights." When he returned to the capital, he found workers still on the picket lines. There was no telegram from the Latvian, no telegram from anyone. Mio sat in his hotel room waiting for a knock on the door and an envelope to be slid underneath. The mission was beginning to wear on him. He recognized the great historical necessity of removing Cukurs from the face of the earth, but personally he'd had enough of the Late One. "During these five months, this despicable man had taken a central place in my life," he recalled. In addition to being a mass murderer of his people, Cukurs had begun to get under his skin. His ratlike suspicions! His groveling! His lusting after the good life! The movie camera trick had been clever; the man was an exasperating nemesis.

The agent sat in the Café Aguila, watched tango dancers in front of the Galerías Pacífico, read newspapers, and spent endless hours walking the streets, burning off nervous energy. As he ambled down the boulevards, he fashioned a settlement to the postal strike in his mind. He thought of his children and his wife. The dull days stretched into a shapeless week.

On February 5, he walked into the lobby, hot and sweaty after another exhausting walk. He checked at reception to see if any telegrams had come in. This time the clerk returned triumphantly bearing an envelope. Mio ripped it open.

> GLAD TO INFORM YOU THAT ALL THE PREPARATIONS ARE COMPLETE. I HAVE THE DOCUMENTS AND THE VISA. WAITING TO HEAR FROM YOU AS TO WHERE AND WHEN I JOIN YOU.
>
> WITH FRIENDSHIP, YOURS, HERBERT

Wonderful news. He grabbed a pen and quickly wrote a telegram addressed to Yariv in Paris.

> NEGOTIATIONS HAVE FINISHED, AND THE TRANSACTION IS ABOUT TO TAKE PLACE. REQUEST YOU SEND URGENTLY THE TEAM OF EXPERTS IN ORDER TO COMPLETE THE TRANSACTION SUCCESSFULLY. ANTON.

Things appeared to be going swimmingly. But, unknown to Mio, Cukurs was still torn between desire and fear. On February 15, he went to his contact at DOPS, the Brazilian secret service, and met with one of the officers there, Alcides Cintra Bueno Filho. Alcides Cintra had a brooding reputation; his nickname was "the Gravedigger." He was later accused by the families of several Brazilian dissidents of supervising the murders of their loved ones under the military government.

"A European business partner is asking me to travel to Montevideo to meet him," Cukurs told Alcides Cintra. "What do you think, can I travel to Uruguay? Isn't it risky?"

"It's safe to say your enemies haven't forgotten you," Alcides Cintra said. He reminded Cukurs that his department's jurisdiction ended at the Brazilian border.

"I trust the man," Cukurs replied.

"Don't go," Alcides Cintra said. "Here you live in peace because we protect you. But don't forget — the moment you leave Brazil, you aren't protected anymore. You expose yourself to your enemies."

Cukurs was even more suspicious than Mio believed. While the Mossad agents made their final preparations, the Late One was thinking of calling off the trip.

TWENTY-TWO

"To Live With a Few Murderers"

THE DEBATE IN THE GERMAN PARLIAMENT was scheduled for March 10. West Germany was in a froth over the looming discussion; that spring, the issue was receiving "almost unparalleled" coverage in magazines and newspapers. One of the major German television networks announced it would televise the proceedings live from the Bundestag. Opinion polls continued to show that, though the numbers had dropped slightly, most West Germans still supported the statute.

The government's months-long investigation into Nazi criminals and cases came to a close and Justice Minister Bucher called a press conference in Bonn to present the results. Bucher stressed the lengths to which the government had gone in hunting down perpetrators. They'd spent twenty years pursuing camp commandants and guards, SS officers and Gestapo killers. Along with the Allied forces, they'd convicted thousands of Germans "due to allegations of war crimes or other National-Socialist offenses." He then came to the results. Despite the government's best efforts, he said, "in individual cases, it is not impossible that further investigations will bring new allegations." Archives, particularly in the Soviet Union, could contain material the Germans had never seen and that material might include evidence of previously unreported mass murders. "Therefore, we cannot exclude the possibility that unknown crimes and unknown perpetrators in leading positions will come to the attention of the public after May 8, 1965." If the statute went into effect, those perpetrators would go unpunished forever.

The Central Office, which was responsible for the prosecution of Nazi crimes, filed its report a few days later. After examining archives in Poland, its officials had found that they needed to open two hundred to three hundred investigations into cases of wartime murder immediately. That number didn't represent the final tally, as new investigations almost inevitably turned up new suspects. Confronted with this fresh material, the Central Office admitted it wouldn't be able to finish reviewing the archival documents, let alone filing the cases, before the statute was imposed.

The findings were a vindication for Tuviah Friedman and others who'd been arguing that hundreds, if not thousands, of unindicted Nazi criminals remained at large. But Bucher refused to concede an inch. "We must be prepared, if necessary," the minister told reporters, "to live with a few murderers." There would be no change. On midnight of May 8, Nazi murderers not yet indicted for their crimes would be granted full clemency.

The news made headlines around the world. The American Jewish Congress called it "a shocking reminder that the moral redemption and rehabilitation of the German people is yet to be achieved." In China, Mao's government branded the decisions "sacrilege to millions and millions of victims," and the archbishop of Boston thundered that "the most monstrous event in our contemporary history must be answered in justice as long as our generation is alive." Fifty-three British MPs signed a petition demanding that the statute be extended. But the Germans resisted. Chancellor Erhard, facing an election later that year, "had to fight for every vote, even neo-Nazi ones."

Leaders in Europe and the United States offered to open their classified archives and send Berlin secret documents on crimes committed by Nazi officers and bureaucrats. The Germans declined, saying that "it is very unlikely" that there were more Nazi criminals to be discovered there. Nahum Goldmann, the founder and president of the World Jewish Congress, volunteered to travel to East Berlin, retrieve the secret files, and hand-deliver them to German prosecutors. He was told to stay home.

The chancellor's office had made its decision and was sticking to it. The Bundestag now became the only hope for Friedman and his fellow campaigners.

The four Mossad team members flew to Buenos Aires separately and checked into their rooms, each in a different hotel. That evening, Mio met the others at a café, where they requested a table off to the side, to avoid attracting attention. Mio briefed them on what had happened at the transit lounge: the movie camera, Cukurs' failure to get the documents, his unending suspicions. At the risk of boring the men, he felt it necessary to warn them one last time about the Late One. "He's like a wild animal," he said in Hebrew. "The minute that he senses he is trapped, he'll muster all his strength."

Before leaving Paris, the others had finished their training with Lichtenfeld. The plan was for three of them to grab the Late One as he entered the house Mio chose for the execution; the last agent, approaching from the back, would knock him to the ground with a special Krav Maga blow. The men talked over the plan and made a few final adjustments. Mio warned them again that Cukurs would likely be carrying a gun. How complete a humiliation would it be if Cukurs managed to overpower four Mossad agents, the pride of the nation, and execute them? It was too horrible to think of.

Mio had one final errand to run. He had a peculiar condition that might affect the mission: he tended to perspire rather heavily. If he met Cukurs in Montevideo, the sweat on his bald pate might be attributed to the hot sun, but it might, in the mind of a paranoid person, also be taken as a sign of nervousness. Mio went to a nearby drugstore and asked the pharmacist if he could recommend anything. The man wrote him a prescription for the antianxiety drug Librium, which had just come on the South American market. Mio tested it the next morning and found that it cured the problem.

The five members went their separate ways and took different flights to Montevideo. Mio arrived on February 10 and checked into the Victoria Plaza Hotel in the city's Independence Square. Eliezer Sudit arrived in Uruguay soon after, checked into a luxury hotel, went to a rental car agency, and booked a green VW Beetle. The others would be needing cars, too, and as Sudit inspected the agency's fleet, he realized they had a problem. Most of the vehicles were in terrible shape: worn-out brake pads, unreliable engines, bald tires. He tried another agency, with the same result. The tourist season was waning, and no maintenance seemed to have been done on the cars since the snowbirds started leaving. Mio finally hired a mechanic to fix the team's cars.

The others flew in one by one and began driving around the city to familiarize themselves with the highways and main roads. They noted landmarks and traffic lights on their maps and timed the drives to various points in the city. They found out-of-the-way nooks where the team members could gather before or after the execution without becoming conspicuous. They wanted to know the city as well as a native so that when they were under stress, the escape routes would come to them effortlessly. This was the grunt work of espionage, "quite gray, exhausting and more often than not also boring," as Mio put it. The team couldn't even push one of the cars over the speed limit for a bit of fun. Yariv told them that the worst thing they could do was come to the attention of local law enforcement; even parking tickets must be avoided.

Mio, who was perhaps more paranoid than the rest, even read the Uruguayan papers every morning, though he spoke no Spanish; the Latin he'd picked up as a young boy in Germany helped him decipher the headlines. He was looking for anything that might tie up traffic during the operation. A visiting head of state, a major soccer game, construction. Day after day, there was nothing.

Sudit spent his time following real estate agents around potential properties. His requirements were quite strict: the place had to be respectable-looking, fit to be the temporary headquarters of a prosperous Austrian firm. It needed to be close to major thoroughfares that would allow the team members to leave the neighborhood quickly without getting tangled in a warren of out-of-the-way streets. And it had to be surrounded by a large garden. Close neighbors might report the shouts of a man in distress. To explain his strict criteria, Sudit told the realtors that he wanted a quiet spot where he could relax with his wife, whom he planned to bring to Uruguay as soon as the house was rented.

Every morning, Sudit and the real estate agents went from neighborhood to neighborhood, looking at prospects. But tiny Uruguay was barely on the tourist map in 1965, and few visitors from Europe or other parts of South America wanted to rent an entire house. They all stayed at the big hotels downtown. The economy was struggling and didn't pull in many foreign nationals or businessmen from other parts of the country looking for residences for a month or two. Turnover was minuscule. For all of Mio's research, he hadn't realized just how small the short-term rental market really was.

Nothing was acceptable. The houses were either too small, too poverty-stricken, or locked in a nest of streets and alleyways you would need a map to get out of. "The days passed and I became concerned," Mio said. Perhaps choosing this obscure country had been a mistake; in Rio or Buenos Aires, they would probably have found a house in a few hours. Now, however, they were stuck with Montevideo.

Finally, Sudit had a stroke of luck. "I met a Greek guy!" he told Mio. When Sudit explained the requirements, the Greek realtor had promised he would find him what he was looking for. The two of them went out the next day to the first house the Greek had found. Sudit studied it as they pulled up: it was enormous, two stories high, too conspicuous. Sudit sighed; this wasn't it. He told the Greek to keep looking.

He did. Soon after, he called and asked to meet. When Sudit arrived, the Greek greeted him with the words "I have just the place for you!" The way he described the house, Sudit also became excited; it seemed to satisfy every requirement. They drove to the address. As they approached, Sudit's spirits rose. From the street, the place looked very presentable, and it came with a nice-size garden.

They rang the bell. A man opened the door. He had long black ringlets cascading down on each side of his face and was wearing a white shirt and black suit. From the top of his curly hair peeked a yarmulke. The man nodded, stepped aside, and gestured for the Greek and his client to enter. Sudit said nothing, only stared. He turned away without a word and walked toward the car.

The Greek was mortified; he blurted out something to the Orthodox Jewish owner and came running after his client. "He was going crazy," Sudit's son recalled. "'*What are you doing? It's perfect for you! It had everything you want!*'"

Sudit got into the car. "I don't like Jews," he said.

The Greek was beside himself; he'd found a diamond in the rough, but his client had turned out to be an anti-Semite. Sudit said nothing. Better that the Orthodox Jew think he'd been turned down because of his yarmulke than that some Fascists firebomb his family for taking part in the Cukurs operation.

Sudit cherished the lie—"his little joke to himself"—but still the team had no house. The date of the Bundestag debate was approaching. The operative returned to the hotel room in a gloomy mood.

The Greek didn't give up. Another morning, he took Sudit to a house called Casa Cubertini, located in the southeastern barrio known as Carrasco. The neighborhood had once been a resort for the wealthy, and its placid, tree-shaded streets that spilled out to the white sand beach of Playa Carrasco still drew the Uruguayan elite. It would be an elegant setting for the final act.

The house itself, however, was unpromising. "Classic it was not," Mio later said. The street it sat on, Calle Colombia (Colombia Street), was in reality a dirt road—hardly the standard for rich

Austrian firms—and the structure itself looked almost abandoned: neglected bushes and ratty grass, bare windows without shades or blinds. When the broker pushed open the front door, they stepped into a living room "covered in a tasteless mosaic of pink and black tiles." The beach was only about a hundred yards away, and the crashing surf could be heard through the open windows. But there was also the sound of hammers and handsaws in the near distance: a construction crew next door was rehabbing a house. They were likely to still be there when the men brought the Latvian by, and they might alert the police about the strange visitors. The lack of shrubbery in front meant that the laborers could see anyone entering and leaving the place.

But the main drawback was the house's ambience. If the place seemed weird or inappropriate even in the small details, it could set off triggers deep in Cukurs' twitchy brain. What would the Mossad team do if he stopped at the entrance and refused to enter? Shoot him and run?

Sudit was getting jittery. "So what do you think?" he asked when he brought Mio to see the house. Mio knew Cukurs; the decision was his. It made Mio jumpy.

The agent ran possible scenarios in his head. If he brought Cukurs to the house, then stood aside at the entrance and said, "After you," the Latvian might balk. "He'll never agree to be the first to enter a house he's never set foot in." That would not do. He couldn't leave Cukurs any time to think; he had to make things automatic. If he got out of the car before Cukurs and strode to the entrance himself, opened the door, and walked in as if it was another in an interminable series of empty houses they'd inspected, Cukurs would be lulled into following. Mio had been training Cukurs' mind, snipping psychological trip wires as it were, ever since São Paulo.

"I'm sure he'll come into the house without a problem," Mio said. "What happens once he's inside and the door is closed behind him is entirely up to you." After consulting with Yariv, Sudit signed a two-month lease, paying the owner $5,000.

The house had been the final piece in the mission plan. Yariv now decided on a date: February 23. (He'd been born in 1923, and considered the last two numbers to be lucky.) "Let it be with luck!" he told the others. Mio telegrammed Cukurs: BUSINESS RUNNING WELL. YOUR HELP NEEDED. I'LL BE GLAD IF YOU COULD JOIN ME IN MONTEVIDEO ON THE MORNING OF 23 FEBRUARY.

The next day, the receptionist handed Mio the reply: DEAR ANTON, ARRIVING BY VARIG FLIGHT ON THE 23RD. This was followed by another, in which Cukurs corrected the flight information. He would be on Air France Flight 83. Mio booked the Latvian a room at his hotel.

Now that the plan was in place, the men had little to do but eat, sleep, and worry. They drove the route to Casa Cubertini countless times, checking their watches as they hit various landmarks. They went out at different times of the day, studying the traffic patterns. The Israeli military motto—"Hard in training, easy in battle"—carried over into the field.

Mio continued to read the newspapers every morning. One item that made news services around the world on February 20 would surely have caught his eye. Papers from Dublin to Washington reported that the statute of limitations controversy in Germany had caused an incident—rather amusing, in one way, and absolutely unamusing in another. A German journalist had written to the Justice Ministry with a simple question: If the statute went into effect on May 8, would that mean that Hitler, provided he'd survived the air raids and suicides in his bunker in 1945, was absolved of all his crimes and could reappear in Berlin as a free citizen? The officers were horrified. *Would it?* they asked one another. The statute would, after all, absolve the Führer of all the millions of murders carried out in his name during the war. The officers of the court consulted experts, some of whom said that Hitler could still be charged with *political* crimes even if he was cleared of criminal charges. But the remote possibility that a stooped, gray-haired Führer could stroll slowly along the Alexanderplatz on some spring afternoon was too much for the German imagination. "We were impelled by the idea: what if Hitler suddenly shows up," said one official, who "added that this was generally regarded as absolutely impossible." The court swung into action. It opened "preliminary proceedings" against the German leader and even requested an arrest warrant in his name, which they told reporters should reassure anyone concerned that, as the *Atlanta Constitution* put it, "if Adolf Hitler returns from the dead, he will have to answer for his crimes." The story ran all over the world with headlines like HITLER FACES COURT — IF HE'S ALIVE and MACABRE MOVE TO PROSECUTE HITLER.

Over breakfast one day, Mio was scanning the local newspapers when he spotted another article, this one about road construction. The main street the team planned to use after the operation was to be closed on the twenty-third for repair. This meant they would probably be stuck in traffic or detoured into a maze of back streets. The men jumped in their cars and scouted out an alternate route back to the city from Carrasco.

The team discussed what would happen with the Late One's remains after the execution. Perhaps just leaving him sprawled on the floor of Casa Cubertini was too informal, too undignified, as if he'd been killed by a home intruder. And the team couldn't very well go shopping for coffins in Montevideo without telling the sellers that the body wasn't quite ready yet. Eventually they came up with a solution: Sudit was tasked with finding a suitcase large enough to fit the body. "He was looking for a large traveling trunk," Mio said, "the kind that features in travel books from the nineteenth century, the kind Thomas Cook passengers had taken along, packing therein half of their household."

Sudit bought a tape measure and went from store to store, bending down to stretch the tape against the leather flanks of various trunks. "The customer was quite demanding," said one salesman. "He measured every trunk — length, width and depth." Sudit stood contemplating the various offerings for twenty minutes before asking the clerk if they had anything bigger. If he was told no, he would move on to the next place. Nothing was large enough to fit the muscle-bound Latvian.

At last, Sudit came to one of the last travel shops, near the American embassy. He went through his routine before finding a leather trunk that fit his requirements. He bought the item, looked at some rugs, and chose two of them. The trunk was far too big for the VW, so the salesman called a delivery service, which brought it and the two carpets to Casa Cubertini.

Mio paid the hotel bill in advance and drove to the car rental agency to settle up with them. He bought two tickets on Lufthansa from Montevideo to Santiago de Chile, a trip that, if everything went according to plan, would never take place. Despite the precautions, Yariv didn't feel comfortable; he felt one crucial asset hadn't been thought of. "We are missing someone who really knows the area well, the mentality and the language," he told Mio. It was too late to summon a native Uruguayan from Israel, if such a person could even be found. So Yariv looked up an old contact now living in the city. He agreed to help them with any local complications and also serve as a lookout on the twenty-third.

A few days before the target date, the men separately made their way to a restaurant. It was the same place Mio had brought Cukurs on their exploratory trip. The maître d' brought Mio to a table, the one he'd had before. When Moti Kfir, the former Sorbonne student, joined him, Mio smiled. "By the way, you're sitting on the very same seat the Late One sat on." Kfir jumped up "as if he had been stung by a bee." He suggested they move to another table.

The others arrived and ordered their meals, then began going over some last-minute adjustments to the plan, keeping their voices low amid the clatter of cutlery and conversation. They were deep into the details when their waiter appeared at the table. "Shalom, friends!" he said in Hebrew. Mio immediately caught his accent: Hungarian. The waiter had overheard the five agents talking and was eager to reminisce with his fellow Jews. "I served with the fourth battalion of the Palmach, and was stationed in Nahariya," he said. The five men looked at one another and burst out laughing. Despite their elaborate precautions, they'd been outed by a Hungarian waiter. They lowered their voices further.

As the day approached, Mio grew tense; he spoke even less than usual. The debate in the Bundestag was only three weeks away. If they failed, another kill team might not be organized in time to meet the deadline. To relieve the stress, he and Yariv decided to go to a casino and play some roulette. After getting his chips, Mio walked over to the croupier. The agent chunked the chips together in his palm and placed them on 23 red.

"*Nada mas, rien ne va plus,*" the croupier called out. The silver ball spun along the groove in the lacquered wood; the two agents watched it hug the perimeter before slowly losing speed and dropping toward the felt-covered slots.

"*Veintitrés!*" the man called out. Twenty-three.

As the croupier paid out his chips, Mio nodded to Yariv.

In São Paulo, Cukurs came to a decision about the trip to Uruguay. "I was always a brave man," he told his contact at DOPS. "I am not afraid. I know how to defend my life. I always carry a gun — and believe me, in spite of all the years that have passed, I am still a fine shot."

The way to his dreams led through Montevideo. He would go.

TWENTY-THREE

The House on Colombia Street

CUKURS WAS SCHEDULED TO ARRIVE on Air France Flight 83 at 9:30 a.m. The afternoon before, Mio, using his raw Spanish, checked the weather reports in the newspaper and saw that a clear, hot day was promised. Delays seemed unlikely. Mio set his alarm for early the next morning. One of the newspapers for sale in the hotel lobby that morning carried the headline BONN GOVERNMENT WILL DISCUSS TOMORROW THE INVESTIGATION OF NAZI CRIMES. It referred to a preliminary meeting before the coming debate. Focused on the day's task, Mio passed the hotel newsstand without noticing the headline.

Mio left the hotel and drove slowly through the waking streets of Montevideo; in the trunk of his rented black VW sat an "escape suitcase" neatly packed with his spare clothes. Dressed in his customary black suit, he steered the Beetle through the light traffic, careful not to speed or turn without signaling. The sun was already beginning to warm the faces of the stone buildings.

He arrived at the airport early and made his way to the visitors' gallery, which was crowded with people awaiting the arrival of cousins, aunts, and friends. As he watched the clock, he occasionally glanced at the opposite end of the hall. After a moment, he spotted Yariv, his profile studiously turned away from his fellow agent. When the Late One arrived, Yariv would memorize the details of his clothing, then quickly depart to pass on the information to the other Mossad men.

In the distance, the tires of the Air France plane threw up a puff of gray-blue smoke as they squelched onto the tarmac. The aircraft rolled over to the reception area, and a mobile stairway was maneuvered toward its sealed door. Once it opened, Mio watched the passengers depart down the gangway. Small groups erupted in cheers when a familiar face was spotted. Businessmen, students, returning emigrants. Mio studied each person as they descended. Finally, he saw Cukurs shoulder his way through the plane's doorway and emerge into the warm Uruguayan sunlight. Once again he was struck by just how strong, how *vital,* Cukurs was, even at sixty-four. He still moved with the rough elegance of a former athlete, every inch the strutting aviator he'd been thirty years before, a function not only of his muscled torso but also of his immense self-regard.

The Latvian peered at the waiting crowd through his glasses and picked out the round, slightly pudgy face of Mio. His face brightened, and his right arm came up in a wave. Two fingers formed into a "V for victory" sign. Even now on his right hip, beneath the outline of his cheap suit, Mio could make out the shape of the 6.35 mm Beretta pistol. As Cukurs came down the stairs, Mio caught Yariv studying the target closely. Then the mission leader turned briskly and slipped through the crowd.

Mio waved back to Cukurs, who stepped off the gangway and strode toward him. The agent stared at the Late One. In their long months together, even as Mio had won the trust of the Latvian, he'd never managed to penetrate into the depths of the man's being. He'd eaten in the man's home, met his family, almost shot him out of sheer rage. But the motivation for his murder spree in Latvia had remained a mystery. Mio had simply assumed he was deeply anti-Semitic.

Cukurs came striding up.

"Good day, my friend," Mio said, smiling warmly.

"I'm so glad to see you," Cukurs said, shaking Mio's hand. "Tell me what's happening and what's going to happen."

"Oh," Mio said, "the best. We have great plans, and you're part of them."

They got into the black Beetle, and Mio headed toward the hotel. "I drove in the most natural way possible, not wanting to give Cukurs the slightest reason to be tense or concerned," he recalled. Mio was pleased; Cukurs seemed to have relaxed since their first trip to Montevideo. They arrived at the hotel at 2:25 p.m., and Cukurs went up to his room. Leaving his suitcase unpacked, he came back down forty-five minutes later and found Mio waiting by the car, double-parked in front of the hotel. Mio had tipped the porter previously to make sure he didn't get a ticket. "What a beautiful room," the Latvian said.

The two men strode into the Lufthansa office next door. Mio wanted to book the tickets to Chile; he spoke loudly enough for Cukurs to hear. The clerk, when interviewed later, said that the two men "appeared to be friends." Once the transaction was complete, the pair got in the Beetle.

"That's settled, then," Mio said as he threaded his way through the light traffic. "Now, to work. I've already found a place that will serve as our temporary office. I'll show it to you later." He groused that the house wasn't up to his usual standards and that he'd have to find another before long. But Casa Cubertini would do for now.

They went to the real estate office, met with an agent, and drove to the Carrasco neighborhood. They were close to the Casa, but Mio was in no rush; they were scheduled to look at three properties that afternoon. The psychological conditioning, the going in and out of houses until it became a boring routine, had to be reestablished. After touring the third property, the agent suggested another that might suit the Austrian, but Mio begged off, saying he had another appointment. The agent said goodbye.

"I'm almost out of petrol," Mio said as he started up the Beetle. He'd purposely let the gas run low, and now he steered the car into a nearby station. Across the street, a red Beetle was parked near the curb. The lookout Yariv had found at the last minute, the Uruguayan Jew, was sitting inside. He turned to peer out the side window, watching the two men chatting as the car was filled with gas. The plan was on schedule. He put the car in drive and moved off into traffic.

Once the black Beetle was ready, Mio paid. "We're really close to the house I rented temporarily," he said. "Come, I'd like you to see the place."

They got into the VW and Mio turned onto Colombia Street. He spotted the red Beetle far ahead of him.

At the Casa, the four men had undressed down to their underwear. If Mio's reports had been correct, the encounter would be bloody, and they didn't want the evidence of a struggle on their clothes. They waited in the hot, humid room, listening to the construction workers' banter and the fall of their hammers. They checked their watches. Minutes later, they heard the black Beetle rumble down the dirt road.

Mio pulled into the driveway. He saw the construction workers, dressed in white undershirts and torn, dirty pants. He saw the white wooden door of the Casa.

"Here we are," he announced to Cukurs. "This is the house."

The brakes squeaked gently as he brought the car to a stop. The sound of the engine died away as he turned off the ignition, replaced by the sound of the workers' chatter and the soft bump of the ocean waves.

He reached into his pocket and took out the house key as he pushed the car door open. Then he was walking toward the door, not glancing back to see what Cukurs was doing.

This is the moment, Mio thought. *He must follow me.*

In his peripheral vision, he saw Cukurs emerging from the Beetle. The agent inserted the key into the lock of the front door and pushed it open. He walked into the house, stepped behind the door, and gripped the edge of the wood with his hand. On either side of the door were Yariv, Amit, Kfir, and Sudit, stripped to their underwear, their eyes boring into his. It was dim inside, a late-afternoon gloom darkening the interior.

The men listened, struggling to make out the sound of footsteps amid the clatter of hammers from next door. Now they could hear Cukurs on the path. His footsteps approached within a few feet of the door. The four men tensed, sweating in the subtropical heat. The Latvian appeared in the doorway; Yariv and the others pressed their backs against the wall. Cukurs took one step into the room, then another, and paused, his eyes searching the corners for his partner. Mio, with a burst of energy, shoved the door with all his might. It slammed shut with a bang.

The four agents rushed at Cukurs. Three grabbed at the aviator's arms, trying to pin them to his sides, while the last came up behind him, raised his arm high, and brought his fist down with all his might on Cukurs' neck, the spot that Imi Lichtenfeld had drilled into him. As Mio watched, the scenario that he'd obsessed over since meeting Cukurs began to unfold in slow motion. Instead of dropping to the floor, as they'd hoped, Cukurs pushed the Mossad agents away with a bellow. The men fell back.

They rushed him again, and the Latvian thrashed at their hands. He struggled free and pivoted quickly toward the door, took two steps, and grabbed the knob. "He fought like a wild and wounded animal," Mio said. Mio and another agent threw themselves against the wood to keep the door from opening, while the others tried to shove Cukurs into the middle of the room. The Latvian was pulling on the doorknob with tremendous strength. The agents grabbed at his arms, sweating and panting, but they couldn't dislodge him.

Suddenly, Cukurs yelled in German, "Let me speak!" Then again. *"Let me speak!"*

The knob tore out of the wooden frame and fell clattering to the floor. The four men hung on Cukurs as he reached for the gun under his waistband. His strength was beyond anything they could have anticipated. Yariv reached up to Cukurs' face with his hand. As Cukurs roared, the Israeli's finger accidentally slipped into the Latvian's mouth. He bit down and his teeth sank into the meat of the finger, tearing the tip off clean. Yariv screamed in pain and snatched his hand away.

The Israelis had lost control. Covered in sweat, they were unable to wrestle Cukurs down. He shouted as he swung at them and shoved them away. They staggered back from his blows, then ran at him again. His Beretta was snug in its holster. He reached for it once more.

Now one of the agents — they would never reveal which of them it was — spotted a hammer on the floor; it must have been left behind in the vacant house after some recent repairs. He rushed to it and grabbed its handle. Turning, he ran at Cukurs, raising the heavy tool above his head. Two steps later, he slammed the hammer down with tremendous force on the top of the Latvian's skull. Blood sprayed onto the men's faces and clothes; the blow was so powerful that it even sent a geyser of red droplets shooting up to the painted ceiling.

Cukurs, wounded but still upright, staggered backward. The agent raised the hammer again and slammed it onto the crown of his head, striking with such force that the wooden handle snapped. Cukurs dropped heavily to the floor.

The plan had gone badly awry. The Latvian sat at their feet, dazed and silent, his breath labored, his shirt soaked with blood, and the white bone of his skull visible through the torn flesh.

Fearing that his shouts might have alerted the workers next door, the agents proceeded to the next step in the mission. "We wanted him to know," said Mio, "that this entire long affair with Anton Kuenzle and his 'business associates' who came to South America had been designed only to set the stage for the moment of revenge in the name of his innocent victims." The revelation of their true motives was woven into the plan from the first; they wanted Cukurs to feel the same betrayal that so many Latvian Jews had experienced beginning on the night of June 30, when their friends and colleagues had turned against them with bewildering speed. That had been the intent. But the scene in the house on Colombia Street had not gone according to plan. It had devolved into another moment from Riga: the night of November 30 and the clearing of the ghettos, as women and children were pulled out onto the street and beaten with rifle butts, their heads shattered as their blood stained the snow and ice.

Yariv was standing away from Cukurs, holding his hand and "moaning in pain." The sound filled the room as one of the other agents walked to the wall where he'd left his clothes; he bent down and retrieved his gun, which was equipped with a silencer. He gripped it in his hand and walked back to Cukurs. He rested the barrel against the back of the man's head. Cukurs was unrecognizable. Blood gushed over his face; his skull was badly fractured. There was no time to read the verdict now, and the man, sunk in a semiconscious stupor, would probably never understand it. The agent pulled the trigger twice, and the Late One's head slumped forward.

The men stood still for a second. Had the construction workers heard the two shots? Were they coming to investigate? Had police already been called; were patrol cars on the way? The five listened. The workers seemed to still be chatting at the same volume; their voices betrayed no urgency. The agents relaxed slightly.

Sudit opened the front door and walked outside. He turned on the hose in the yard, and the operatives went out in turn to wash the blood off their skin. When they went back inside, the men searched Cukurs' pockets and removed his passport. They slid the Beretta out of its holster and carried the body to the large trunk that Sudit had taken days to find. They lowered Cukurs into it, but the body was too big. They pushed on the Latvian's legs, slowly squeezing the body into the trunk as the leather stretched. One of them laid a sheet of paper on the corpse. It was the verdict they had intended to read before executing the Latvian:

> Considering the gravity of the crimes of which HERBERT CUKURS is accused, notably his personal responsibility in the murder of 30,000

> men, women and children, and considering the terrible cruelty shown by HERBERT CUKURS in carrying out his crimes, we condemn the said CUKURS to death. He was executed on 23 February 1965.

It was signed "Those Who Will Never Forget."

Mio would later talk about the sheet of paper with the verdict, but the team also left another document that went unmentioned for years. It was an eyewitness account that had been read out at the Nuremberg trials by the chief British prosecutor, Sir Hartley Shawcross, in late 1946. It described an *aktion* that had taken place near the city of Dubno, on the Ikva River, in present-day Ukraine. On October 5, 1942, a German engineer named Hermann Friedrich Graebe had gone to Dubno to visit a site where some grain storage buildings were being built; it was part of his job to check that the construction was proceeding according to his company's specifications. When he drove out to the location with his foreman, he found that "great mounds of earth" had been excavated from three large pits near the silos. Families of Jews—Graebe could see the bright yellow stars sewn onto their coats—were being taken off idling trucks by members of a Ukrainian militia and ordered to undress, under the eye of an SS officer carrying a dog whip. Graebe and the other man got out of their car and approached over the muddy ground; no one stopped them. As Graebe walked, he came across a large stack of about two thousand shoes, along with "great piles of under-linen and clothing." Close to them were groups of naked Jews. The German engineer gazed at them. What the Mossad agents left on Cukurs' body was the description of those nameless victims and how they'd behaved toward the end:

> Without screaming or weeping these people undressed, stood around in family groups, kissed each other, said farewells, and waited for a sign from another SS man, who stood near the pit, also with a whip in his hand. During the 15 minutes that I stood near I heard no complaint or plea for mercy. I watched a family of about 8 persons, a man

> and a woman, both about 50 with their children of about 1, 8 and 10, and 2 grown-up daughters of about 20–24. An old woman with snow-white hair was holding the 1-year-old child in her arms and singing to it and tickling it. The child was cooing with delight. The couple were looking on with tears in their eyes. The father was holding the hand of a boy about 10 years old and speaking to him softly; the boy was fighting his tears. The father pointed to the sky, stroked his head and seemed to explain something to him. At that moment the SS man at the pit shouted something to his comrade. The latter counted off about 20 persons and instructed them to go behind the earth mound. Among them was the family which I have mentioned. I well remember a girl, slim, and with black hair who, as she passed close to me, pointed to herself and said "23." I walked around the mound and found myself confronted by a tremendous grave. People were closely wedged together and lying on top of each other so that only their heads were visible. Nearly all had blood running over their shoulders from their heads. Some of the people shot were still moving. Some were lifting their arms and turning their heads to show that they were still alive. The pit was already two thirds full. I estimated that it already contained about 1000 people. I looked for the man who did the shooting. He was an SS man, who sat on the edge of the narrow end of the pit, his feet dangling into the pit. He had a tommy gun on his knees and was smoking a cigarette. The people, completely naked, went down some steps which were cut into the clay wall of the pit and clambered over the heads of the people lying there, to the place to which the SS man directed them. They lay down in front of the dead or injured people; some caressed those who were still alive and spoke to them in a low voice. Then I heard a series of shots. I looked into the pit and saw that the bodies were twitching or the heads lying motionless on top of the bodies, which lay before them. Blood was running away from their necks.

The men lay the sheets with Graebe's testimony on the body in a folder with the verdict and closed the trunk. "We clean and we vanish," one of the agents said. They took rags and began wiping down the walls and the floors; none of them apparently glanced up to see the streaks of blood on the ceiling. They moved on to the doorknobs, which they wiped clean of fingerprints, then did the same to the trunk. When they'd finished, they dressed in different clothes than they'd worn earlier and left the house.

Yariv and Mio drove away first in the black Beetle. The others checked the grounds one last time for anything they'd left, then got in the green Beetle and followed.

TWENTY-FOUR

The Wait

IN THE BLACK BEETLE, Yariv was in pain; his finger was beginning to swell. "Can you go a little faster?" he asked Mio, who was still wary of exceeding the speed limit. Mio dropped Yariv at the Plaza de los Treinta y Tres near a café and went looking for a place to dump the car. He found a spot, pulled over, parked the Beetle, wiped the steering wheel and gear shift clean of fingerprints, along with the door handle, and retrieved his suitcase from the trunk. Then he walked away from the car and hailed a taxi back to the café.

The other team members gathered there soon after. In their new clothes, they looked like tourists or perhaps middle managers at one of the local offices having a drink after work. "Not one of the café's patrons could have imagined that half an hour earlier these five men had killed a sadistic Nazi criminal," Mio said. He was so focused on finishing the mission and getting out of Uruguay safely that the full import of what he had done — the death struggle, the mysterious cry of "Let me speak!" — hadn't yet fully registered with him.

The five ordered drinks, their faces revealing nothing. "We remained absolutely calm and composed," Mio said, "at one with ourselves and our deed, without even a twinge of a guilty conscience . . . We felt proud of having had the privilege of taking part in this operation." His only regret was that there weren't more missions like this on Mossad's calendar. Cukurs would have to stand in for so many other escaped criminals as the parliamentarians in Bonn began their debate.

Yariv congratulated the men on performing their duty and wished them well on their return home. They got up and left separately. Mio went to a pay phone and dialed the number for the Royal Plaza. He told the receptionist that he and Mr. Cukurs wouldn't be returning to the hotel that night; something had come up, and they had to leave the city. The bill was paid in full; the clerk thanked him for alerting the hotel. Mio dropped another coin in the slot and called Lufthansa to cancel the plane reservations to Chile. He then picked up his suitcase, hailed a taxi, and headed to the airport for his flight to Buenos Aires.

From the Argentinian capital, Mio wrote a letter to the Late One, addressed to his home in São Paulo. He wanted to put the family's mind at ease for the moment and prevent any panicked calls to the Brazilian police. "My dear Herbert," he wrote.

> With God's help and the assistance of some of our countrymen I made it to Chile in one piece. I'm taking a little rest now after the exhausting journey. I'm sure you too will make your way home very soon.
>
> In the meantime, I found out that we were followed by two strangers, a man and a woman. We must be more cautious and alert. As I told you over and over again, you're taking a great risk by living and operating under your real name. Such a risk could be disastrous for us, and also expose my true identity.
>
> Anyway, I just hope that the complications we encountered in Uruguay will serve as a lesson for the future, and that from now on you'll be more careful.
>
> If you detect anything suspicious in your surroundings or your house, do remember my advice — disappear for a year or two among Von Leers' men, until this debate over the Statute of Limitations for Nazi war crimes dies out.

> The minute you get this letter, please send a reply to Santiago de Chile, to the address you already know.
>
> Yours, Anton K.

He sent the letter to a Mossad contact in Santiago, who placed a Chilean stamp on it and dropped it in a postbox.

The note dripped with false leads. "Von Leers' men" referred to Johann von Leers, an anti-Semitic former professor and author of books such as *Juden sehen dich an* (Jews are looking at you). He was currently living in Egypt, where he'd converted to Islam and was serving as the leader of President Nasser's propaganda department. When Cukurs failed to return home, the letter would indicate that he'd disappeared into some kind of underground railroad for ex-Nazis run by von Leers. The other cues would paint Kuenzle as an escapee from justice and insinuate that unnamed dark powers were in pursuit of both him and Cukurs. It would take weeks for Cukurs' family to puzzle them out.

Mio and Sudit returned separately to Paris. There Mio wrapped up the mission's affairs, retrieving the money he'd deposited at Credit Suisse in Zurich. He took the cash and walked out of the bank, nodding to Sudit, who'd stood guard outside in case there were any complications. The last record of Anton Kuenzle had been erased. "As I left the bank, the Austrian businessman who had come to life only six months before, had an address and identity and had spent months in South America, ceased to exist." The men took their families out to dinner, played with their children, made love to their wives. They savored the knowledge that Cukurs' execution would soon be known the world over. The other team members flew to Tel Aviv, where Yariv—in terrible pain—was rushed to Tel Hashomer hospital, where surgeons worked to save his mangled finger.

One agent was given the task of announcing the Late One's fate to the press. He called a list of major German news outlets and read out a statement asserting that the war criminal Herbert Cukurs had been killed in Montevideo by a group calling itself "Those Who Will

Never Forget," to avenge the callous murder of millions of Jews in Europe. When he finished his prepared speech, he would hang up before the reporters could voice their reactions.

They "waited with bated breath." Mio unfolded his newspaper every morning and read it front to back, looking for Montevideo bylines. He tuned in to the evening television news and watched until the announcers switched to the sports results. But day after day passed, and there were no stories from Montevideo, no photos of Herbert Cukurs, no headlines to influence the members of the Bundestag. It was as if the past six months had been a dream. "We became concerned. What could have happened? How come no one was sent to check the story?"

Another week went by and still nothing. The debate in the Bonn parliament was only days away. Mio and the others fretted. It began to dawn on them that their efforts to spread the news may have been misguided. As dwellers in the secret and secretive world of espionage, they'd been trained to avoid even a hint of publicity. "Mio hated journalists," said his son, and his hatred had bred ignorance. The five members of the team had no idea how newspapers or television stations actually worked. Would a reporter fielding an anonymous call in Berlin about a murder in Montevideo, seven thousand miles away, actually feel compelled to investigate the tip? Or would he return to covering the results of the latest municipal elections? Mio and the others were bungling the mission at the last moment; the silence mocked them. "We had not taken into account," Mio said, "the fact that news agencies are commonly swamped with the weirdest callers promising to expose the strangest and most sensational stories — from encounters with UFOs over Berlin to Nazi abductions in Antarctica." In addition, the Nazi-hunting market was overinflated at that particular moment. The statute controversy meant that Nazis were being spotted everywhere, from Bolivia to Gdańsk; few of those tips led to legitimate cases, making journalists wary of spending even a few minutes checking on them.

The corpse of Herbert Cukurs was decomposing in the sweltering heat of a posh suburb of Montevideo, but Mossad had no way of convincing anyone of it. Now the letter to Cukurs' family looked unwise. If Milda Cukurs had called the Brazilian police and reported her husband missing, it might have triggered an investigation. But Mio had outsmarted himself. No one was looking for Cukurs.

"We have no choice," Yariv said after a few days. "We must lead them to the house in Montevideo."

The team drew up a document that included a detailed history of Cukurs' actions in Latvia and the facts of the execution. "His body can be found at Casa Cubertini, Colombia Street, Montevideo, Uruguay," it went on. "Please excuse the form of the letter. We have chosen this method for security reasons; surely you understand." They sent the note to news agencies in Bonn, Frankfurt, and Düsseldorf. But the team's luck still ran bad. On receiving the letter, the United Press International office threw it in the garbage, thinking it was a joke.

By March 5, ten days after the execution, there still hadn't been a single news report. The German legislators would begin discussing the statute in five days. That afternoon, a phone rang at the news desk of Reuters' Bonn bureau.

"Give me your office manager please," a man told the reporter who answered.

The manager came to the phone. He asked the man's name and why he was calling.

"My name is Schmidt. I would like to tell you about a very urgent matter. Did you get the letter on the Latvian case?"

The manager was unfamiliar with any Latvian case. He told the caller he was going to put him on hold and ask his reporters. The man immediately hung up.

UPI was next. This time the man called himself "Mr. Muller." He asked if the bureau had contacted anyone in Montevideo. Then he ended the call. The agencies reported the calls to Interpol, and at least one of them made a call to police headquarters in Montevideo. They asked about reports of a dead body at a place called Casa Cubertini.

The next day, a patrol car drove through the Uruguayan capital and pulled up in front of the Casa. Two policemen got out and approached the house; they noticed that the lock on the front door had a key broken off inside, which would prevent anyone from turning the mechanism. They knocked on the white wooden door. No answer. The policemen brought their hands to their faces; some horrible stench was seeping out from the gap underneath the door.

The men pulled their guns out of their holsters and broke the glass on a side window, then stepped through. The odor was overwhelming now, and the officers noticed dark crimson spots on the walls and floors, blood spatter that the Mossad team had missed. When they walked into the living room, they saw a locked trunk, along with two .22-caliber shell casings and fragments of a gun lying on the floor. They broke the lock on the trunk and raised the lid. There, beneath an immaculate bright red folder containing a number of papers, was Herbert Cukurs.

TWENTY-FIVE

An Offer

THE STORY FLASHED AROUND THE WORLD. There were headlines in Tel Aviv, Moscow, Buenos Aires, Paris, and New York: THE KILLER OF LATVIAN JEWS HAS BEEN ELIMINATED read one; another paper described in detail what had happened to "the galloping Angel of Death." The press would continue to follow closely every twist and turn in the case for months to come. News agencies the world over ordered their local freelancers onto the case and asked them to find out who "Those Who Will Never Forget" actually were. Some newspapers flew their own correspondents into Montevideo. The American journalist Jack Anderson, who'd interviewed Cukurs five years before, wrote a sequel titled "The End of a Nazi." "Scattered throughout Latin America today," he said, "are men who live in terror, Nazi war criminals who got away but now feel the hot breath of vengeance on their necks."

Word reached the Cukurses, who were deeply pained by the death of their father and husband; the family announced to the newspapers that "Jewish terrorists" had killed an innocent man. "The death of Herbert Cukurs," said his daughter, Antinea, "was the death of a hero." In Montevideo, swastikas appeared on walls and telephone boxes, and neo-Nazis attacked a synagogue, throwing a bomb and firing a Colt .45 pistol through its windows. Jewish stores, families, and clubs reported getting telephone threats, and several requested police protection. A letter was sent to the Israeli ambassador to Uruguay. "Mr. Ambassador," it read, "a trunk has also been prepared for you. Assassin!"

In Las Vegas, Sasha Semenoff (formerly Abram Shapiro) was working as a bandleader. The past twenty years of his life had unfolded like a star-spangled dream: he'd played for three presidents and become known as "Frank Sinatra's favorite violinist." When the singer gambled at craps and blackjack, Sasha would be there, playing his violin table-side. Despite his many triumphs, he found that the memory of his persecutors never left him. "I would hear Cukurs laughing and drinking and having fun," he said. "He was horrible. Even though I survived, it's something I can never get over." Semenoff often visited elementary schools to talk about Riga, the Arājs Commando, and the brutality of the Shoah. When he heard the news of the Late One's death, his son remembered, he pumped his fist in the air. "They got him!"

Journalists began competing to file the latest tidbits: The hammer covered with blood and human hair, the mysterious doorknob, the Omega wristwatch found on the floor, the hands, as in an Agatha Christie novel, stopped at the presumed time of death: 5:10 p.m. Inside the room, the police even found a medal inscribed HERBERT CUKURS, MEMBER OF HONOR, INTERNATIONAL AVIATORS LEAGUE. IN RECOGNITION OF HIS CONTRIBUTION TO THE DEVELOPMENT OF AVIATION. Cukurs had carried his precious award with him until the end.

Conspiracy theories sprouted overnight. There'd been an Israeli ship, *Har Rimon*, anchored in the harbor at the time of Cukurs' death. Did its presence indicate that the killing had been the result of a failed kidnapping? There were rumors that the trunk he was found in had air holes punched in its side. (The trunk, which is now at the National Police School in Montevideo, shows no such holes.) At least one journalist wrote that the operation was a case of "Eichmann, part two."

The police interviewed the owner of Casa Cubertini, Antonio Giménez Vidal, and took the Greek realtor, whose name turned out to be Denis Mavrydis Kalogeropulu, into custody. Mavrydis told them about the Austrian man he knew as Oswald Taussig, who wanted a place to relax with his wife. Mavrydis said he'd shown Taussig several houses and spent a good deal of time with him, until the day Taussig revealed that he had met a woman in Montevideo whom he was seeing. The Greek had taken the hint and hadn't visited the Austrian again. After detaining him for two days, the police could find nothing deceptive in the realtor's testimony and released him.

The focus switched to Anton Kuenzle. The police were able to trace his comings and goings to and from Uruguay and follow his trail in Brazil. He had been a demanding client; one car rental company remembered he'd asked for a Chevrolet with a large trunk and offered to pay double the going rate for it. But that was hardly unusual; mostly they remembered Kuenzle as a businessman with expensive tastes who seemed obsessed by the tourist market in South America.

The correspondents had a field day with the identity of the victim. Though the Cukurs family had identified the body, one Uruguayan newspaper reported that the man inside the trunk wasn't Cukurs at all, but Anton Kuenzle himself. The police dismissed the rumor. The Brazilian journal *A Notícia* ran a long article with a hot new theory: the body was indeed that of Cukurs, they argued, but the "verdict" left on the body was a red herring. In fact, Cukurs had been caught in the middle of masterminding a trap designed to catch Josef Mengele, the Angel of Death, and was planning to turn the doctor over to Mossad in exchange for a pledge that Cukurs would be allowed to live free.

The final wave of protests against the statute began as the debate in the Bundestag approached. In the United States, two Jewish senators introduced a resolution into Congress asking the government to pressure the West Germans to block the amnesty. Thousands of protestors marched through the streets of Tel Aviv on March 7. Simon Wiesenthal's fresh connection to the Kennedys would pay off when RFK attended a meeting of the Jewish Nazi Victims of America. "We must not let the world forget the toll that was taken by man's inhumanity to man just twenty years ago," the newly seated senator from New York told the attendees. "There can never be a time when we are no longer ready to bring the criminals of that awful time to the bar of justice." The NAACP added its name to the list of those calling for the statute to be voted down.

Behind the scenes in Bonn, politicians on both sides of the debate worked long hours to find a compromise. The statute could be changed to thirty years, one legislator proposed, giving prosecutors another decade to ferret out the last *genocidaires.* Or the date the clock began ticking could be moved forward to 1949 instead of 1945, allowing them four more years. But the Bundestag's Justice Committee, led by Adolf Arndt, rejected the bill, citing as one reason the idea that "the perpetrator has become another person" — that is, the killers presumably felt remorse for their actions and had matured away from violent anti-Semitism. Jewish groups fumed, pointing out that many perpetrators still proclaimed their hatred of Jews. It also infuriated them that the six million were not even mentioned in the committee's report.

To break the stalemate, the government made a secret offer to the opposition: they would agree to change the statute if "ordinary SS men" who'd committed crimes during the war were exempted from prosecution. It seemed a reasonable suggestion: many nations forgive the actions of its common soldiers for killings on and off the battlefield. But when the opposition dug into the fine print, they found the amnesty was far more extensive than it appeared. It would, in fact, cover "all perpetrators within the National Socialist administrative machinery." The offer was a whitewash, a blanket reprieve.

Shocked, the opposition refused. There would be no deal before the Bundestag debated.

Television cameras were set up in the Bundeshaus, the former Pedagogical Academy in the Platz der Vereinten Nationen in Bonn. Victims of the Shoah, as well as perpetrators who'd yet to be indicted, waited for the deliberations to begin. "If Germany lifts the burden of legal guilt," wrote the *Jerusalem Post,* "the killers could dance on the graves of their victims."

In Paris, Zeev Sharon, the young son of Eliezer Sudit, was at home in his family's eighth-floor apartment. It was a typical French flat, exceptional only for the fact that his mother allowed nothing manufactured in Germany—no radio, no fancy appliance, no toy—to cross its threshold. The family had just finished dinner. Zeev's father announced that tonight they were all going to watch the evening news together in the living room. "There's going to be a special report," he said. "And I want you to tell me if you recognize anyone in it." Zeev was puzzled. His parents never watched the evening news, certainly not with the boys gathered around them. "What was this?" he thought. The five of them trooped into the living room and sat in front of the black-and-white TV as it buzzed on; the gray-and-white static slowly cleared, and a picture emerged. The music that preceded the newscast began. Zeev looked up at his parents and then back at the TV.

The first item was about the execution of a man in Uruguay. Zeev didn't catch the name, but there was something about two bullets and a note and the Holocaust. The camera cut away from the announcer and showed a sketch of the suspects supposedly involved in the crime. Zeev and his brothers leaned forward and studied the drawing closely. The announcer reappeared and went on to the next item.

His father stood up. "Well, did you recognize anyone?"

Zeev shook his head. "No," he said. His brothers agreed. What was this all about?

Eliezer Sudit breathed an audible sigh.

In Mio's house, a similar test was conducted. Interpol had distributed a sketch of "Anton Kuenzle" under a stark WANTED headline. The Israeli newspaper *Maariv* printed it on the front page. Mio's mother-in-law called her daughter to say the sketch bore a strong resemblance to Mio. Was it him? Had he helped kill Herbert Cukurs? Mio's wife answered no to both questions, but she couldn't get the thought of the flier out of her mind. She found the sketch in a local newspaper and brought it home to her young son, whom she'd always found to be quite perceptive.

"Do you know who this is?"

The boy studied the sketch for a moment.

"Farouk!" he said.

He was referring to His Majesty Farouk I, the former king of Egypt, now in exile in Italy. Mio's wife looked again at the sketch. The face did bear a striking resemblance to the pudgy ex-king, with his dark, hooded eyes. She was relieved. If Mio's own son couldn't recognize him, perhaps they were safe.

TWENTY-SIX

The Legislator

THE DAY OF THE DEBATE, March 10, arrived. Millions of West Germans turned their televisions to the channel broadcasting it, and newspaper correspondents from all over the world stood in line for seats in the visitors' gallery, along with students and ordinary citizens. Below the gallery was the semicircular chamber with its dark wooden seats facing the rostrum; behind the rostrum hung a stylized portrait of a black eagle, the emblem of the German state. The gallery seats had been spoken for weeks before, with fifty requests for every spot, and when the guards opened the doors, the lucky few, among them a squad of buzz-cut army recruits, filed into the rows and took their places.

The main political parties had released their members from following their leaders; the men and women were free to vote as they saw fit. The legislators who began filing in early that morning represented the full spectrum of war experiences. Among them were men who'd served the Third Reich faithfully as members of the Nazi Party, as well as Dr. Eugen Gerstenmaier, the last surviving member of the famous July Plot to kill Hitler. Gerstenmaier had heard the execution squad chambering the rounds into their rifles as Claus von Stauffenberg, one of the leaders of the plot, waited for his death sentence to be carried out. He'd even heard Stauffenberg's last words: "Long live our sacred Germany!"

When the session opened, Justice Minister Bucher stepped to the lectern. Handsome and solemn, he had the look of a fit but aging child, his broad forehead topped by a shock of graying hair. He began by acknowledging the passionate interest that the issue they were about to debate had aroused in Germans and non-Germans alike.

"Mr. President," he said. "Ladies and gentlemen. There is hardly any question that has moved people's minds at home and abroad as much as the question of the statute of limitations on Nazi crimes." Bucher acknowledged that the law had met with a "lack of understanding, fierce criticism and hostile rejection" from people around the world and assured those who had suffered under the Third Reich or who had lost loved ones to the "terror of wicked murder" that they would be heard in the coming hours. "Your voice," he said, "has weight." Bucher stated his reasons for supporting the statute: justice for the six million, in his view, had already been achieved. He again pointed out that thousands of Germans had been sentenced by Allied and German courts in the twenty years since the war had ended; foreign archives had been examined; millions of documents had been scrutinized. "I can hardly believe," Bucher said, "that it is possible to take seriously the assertion made by various parties that in the Federal Republic or in the world there are tens of thousands of National Socialist murderers who will go unpunished." The right side of the chamber, where the conservative members of the ruling Christian Democrats sat, rang with applause. Bucher took his seat.

The debate now began in earnest. A legislator implored the deputies to affirm that "one unpunished murderer among us is one too many . . . If we abandon the hunt for them, we might as well abandon the republic and revert to living in caves." A conservative deputy disagreed sharply. The German people, he argued, shared no collective guilt for what had happened during the war; they weren't Hitler's accomplices but his victims. A former justice minister stood up at the rostrum and declared that the issue was "purely legal" and that the statute must be implemented; to change the law would be to pave the way for worse injustices in the future. The speeches were broken up by calls and cries of "Hear! Hear!" or "Very true!" when a speaker got off an especially rich dig at his opponents. Much of the talk was narrowly legalistic; if a tourist had wandered into the gallery above without any idea of what was happening, he might have thought the voices below were talking about tariffs or something just as bland. "At the core of the legal dispute," said one legislator in a typical line of the day, "is the question of understanding the rule of law today." The word "interpretation" was uttered repeatedly. It was, for most of the morning, a day for attorneys and not historians.

The debate rolled on, with applause and dark mutterings from the floor as each speaker tried to pull legislators to his side. Deputies spoke of the distinction between crimes and "political errors"; another declared that "our duty is to ward off scandal from the German people." The speeches and questions went on for seven hours. One of the last to come to the lectern late that afternoon was Adolf Arndt. A few days earlier, he'd affirmed his support for the statute; now the members watched him rise from his seat and walk down the aisle, expecting him to lay out his reasoning. Arndt made his way toward the front of the hall as the "winter sun lay golden stripes across the heads and shoulders" of the waiting legislators. "He is an ungainly figure . . . ," wrote one foreign correspondent watching from the gallery, "not tidy in dress, awkward until the moment when he arrives at the rostrum and takes hold of the desk." Arndt looked out over the floor, the progressives to the left and the conservatives to the right, and began to speak in his rather pedantic style. He skillfully wound through the legal intricacies of the case. Then he sketched a scenario in which the statute went into effect and another war came to Germany. What precedent were they setting by letting the criminals go free? How would that decision echo forward into history? "Future perpetrators [would] think, 'The more murders we commit and the greater chaos we create, the less time the others will have to prosecute this.'" Opponents of the law applauded. It was clear that something unexpected was happening; this wasn't the speech they'd anticipated. Arndt had switched his position; he was coming out against the statute. "In this hour," he told the legislators, "the heart must speak."

Arndt gripped the edge of the rostrum. He paused, letting his arms rest on the raised sides. And then he began to speak again. "A man who takes an infant by the feet in front of his mother and shatters his head on the nearest iron post, a man who has 20,000 or 30,000 people shot or killed, a man who trains his dog so that he tears apart a prisoner's genitals before the prisoner is put to death in the most cruel way, a man who forces prisoners to kneel in the pit they have dug themselves, then gives them 'the neck shot,' and then the next victim comes in, so that for days a fountain of blood splashes out of this mass grave, one can not say of this man: Why is he still dealing with his act today?"

The portrait of the representative soldier was curiously specific. Arndt had apparently been hearing about Cukurs' crimes in the German and foreign media; the Mossad team had specified the 30,000 Jews the Latvian had helped murder, and now Arndt had repeated that same number. No other war criminal who'd gone on trial in Germany in the past few years had been accused of killing that exact number of Jewish men and women. It was unique to Cukurs. And with the exception of the detail about the trained dogs, the other details he gave had all been prominently mentioned in accounts of Cukurs' life. One can only surmise that the Nazi criminal the legislator was referring to had just endured a very public death in Montevideo.

Arndt looked up at the gallery. "We all knew, really," he said. "In my town of Marklissa, we knew the mental defectives and the cripples were being taken from the hospitals and murdered. One knew. There were enough refugees, men on leave from the occupied East, who told their weeping wives, their mothers, their families, what they had seen in Poland and how they could do nothing to stop it." He told the legislators what he'd once said to a group of young people: if their own mother swore on her deathbed, while holding her hand on a Bible, that she knew nothing about the Shoah, then it could only be because "to have known . . . is too terrible." Everyone knew.

There was a disturbance on the right side of the floor. The conservative legislators muttered loudly in horror. Arndt pushed on. "I know I also share the guilt," he said. "You see, I didn't stand in the street and scream aloud when I saw them driving away our Jews in lorries. I didn't put on the yellow star and say, 'Take me too!' . . . I cannot say that I did enough. I don't know who can say they did." He paused. "We must take upon ourselves this very heavy and, alas, very unpopular burden. We must not turn our back upon the mountain of guilt and sin which lies behind us. Instead, we must come together to be what we really are: small, humble servants of righteousness."

Arndt's words were legitimately shocking. It was practically unheard-of for a prominent German in 1965 to confess that they'd played a part, however small, in the Shoah. "He said something . . . ," the *Jerusalem Post* wrote, "rarely heard in Germany." In stating that "we all knew, really," he was admitting that the pretense of innocence couldn't be maintained any longer. The applause swelled up from the progressives, and Arndt turned away from the lectern and slowly walked back to his seat.

The debate ended with two proposals to extend the statute, a strong indication that the amnesty would be defeated. Procedural delays pushed the vote back; on March 27, legislators were scheduled to cast their ballots on extending prosecutions for another five years. In the intervening weeks, more articles about Cukurs ran in German and international newspapers, with headlines such as WAS CUKURS TO BE KIDNAPPED LIKE EICHMANN? and THE CUKURS AFFAIR IN MONTEVIDEO. The German correspondent for the Associated Press filed a long dispatch from Bonn. "The [Cukurs] case came to light less than two weeks before the debate in the West German Parliament," the journalist wrote. "Could it have been intended to draw attention to the prospect that many Nazi murderers may never be prosecuted?" The question was in the ether.

On March 18, the influential conservative thinker William F. Buckley Jr. wrote a column carried in newspapers across the United States, including ones read closely in European capitals. "I do not know anything at all about Herr Cukurs and under the circumstances am quite prepared to believe that it is true that he was guilty," he wrote. For him, that wasn't the point. "The great question of the day, of course, has to do with whether the German government should extend beyond next May the Statute of Limitations." Buckley was troubled by the note left on Cukurs' body. Should every people wronged by another people form a committee of "Those Who Will Never Forget"? Should Americans remember forever the Japanese torture and execution of their fathers and brothers during World War II? Buckley believed that "private acts of lynch justice, like that which was done in Montevideo," were infinitely preferable to changing the law. Ever the contrarian, he was coming out *for* the statute. It was a minority, incendiary position. But if Minister Bucher and his allies took comfort in the idea that at least *someone* was supporting them, headlines such as the one on Buckley's column, LYNCH JUSTICE IS FAR BETTER THAN TAMPERING WITH JUDICIAL SYSTEMS, hardly burnished the cause.

Bucher wanted a clean debate about the law. Cukurs kept dragging other things — the pits, the digging in the forests — back into view.

March 27 finally arrived. When the ballots were cast, the count was 361 in favor of the extension of Nazi prosecutions, 96 against, and 4 abstentions. The statute was soundly defeated. There would be no amnesty for Nazi murderers.

In Haifa, Tuviah Friedman was overjoyed. He'd been among the first to raise the alarm about the amnesty, and the triumph in Bonn belonged to him as much as it belonged to anyone. He celebrated with his wife and colleagues at the Institute of Documentation and accepted the congratulations of politicians and Jews from around the world. After the capture of Eichmann, Friedman had commented, "All those years, I was a beaten man. But I had patience." Now he'd led a campaign whose victory was arguably deeper and more far-reaching than that mission.

The news for his fellow Nazi-hunter Simon Wiesenthal was even sweeter. The campaign had made him into an éminence grise. "The success of Wiesenthal's public struggle against the statutes of limitations," wrote one biographer, "gave both his work and his status a new dimension. From now on he was no longer seen as an amateur sleuth or as a pest running from one official to another . . . but as a personage to whom many doors were open and who had something to say on fundamental questions." Once again, Friedman had been outmaneuvered by the canny Wiesenthal on the publicity front, but he expressed no bitterness over it. The prize was large enough to share.

Outside Germany, the vote was met with nearly universal approval. An article in the *Jerusalem Post* called it "a victory for the moral conscience of men everywhere." Arndt's speech was celebrated as a rare and effective piece of truth-telling. The legislator reached his pinnacle of fame, and the parliament was recognized as having experienced one of its *Sternstunden,* or "finest moments." "The debate in the Bundestag was one of the great debates," the *Post* reporter wrote. "That much is agreed in West Germany." The article mentioned that those who watched the debate ended the day "feeling a little stronger and cleaner." And the events in Montevideo didn't go unmentioned. "The murder of an old Latvian Nazi named Cukurs . . . ," wrote one American newspaper, "suggest[s] the decision to let the limit run on awhile is as close to right as any could be."

There were detractors. Some Israelis felt that five years was a slap in the face. Others would come to believe that the successful Cukurs mission persuaded the country's leaders that extrajudicial killings were a kind of statecraft and would lead them into a tangled morass of assassinations in the coming decades. For them, Mossad had been *too* successful in Montevideo, and so Israel had learned the wrong lesson.

In Bonn, Justice Minister Bucher announced his resignation immediately after the vote, as promised. His career in the Bundestag was over.

In the years following the debate, hundreds of ex-Nazis were tried for crimes ranging from mass slaughters carried out by *Einsatzgruppen* to individual killings inside Auschwitz. The vote gave new momentum for trials of previously indicted Nazis. The Central Office was originally allowed only eleven lawyers; now that number was increased to fifty. In 1967, the handsome commandant of the death camps Sobibór and Treblinka, Franz Stangl, was found living in Brazil and brought to Germany for trial, where he was accused of the murders of 900,000 people. The inmates at Treblinka had named Stangl "the White Death" for his habit of wearing a fresh white uniform as he strolled the paths of the camp, which he'd ordered lined with flowers. He had at least been honest about his attitude toward the prisoners:

> To tell the truth, one did become used to it . . . They were cargo. I think it started the day I first saw the *Totenlager* [extermination area] in Treblinka. I remember [SS officer Christian] Wirth standing there, next to the pits full of black-blue corpses. It had nothing to do with humanity—it could not have. It was a mass—a mass of rotting flesh. Wirth said, "What shall we do with this garbage?" I think unconsciously that started me thinking of them as cargo . . . I sometimes stood on the wall and saw them in the "tube"—they were naked, packed together, running, being driven with whips.

Stangl died in prison of a heart attack six months after being sentenced to life imprisonment.

Even those who held less important posts in the Third Reich felt the effect of the Cukurs operation. The aviator had been a captain, a tiny cog in an obscure corner of the Nazis' wartime dominion, but he hadn't been too small for the Israelis to target. "[The] execution had a profound effect on Nazi criminals living in hiding all over Latin

America," wrote one intelligence agent, "who were stunned to learn that the long arm of the Mossad could reach a person like Cukurs... Many Nazis cut off their contact with old comrades, fearing betrayal by those who knew all of their past."

The debate and the vote, along with other factors — among them the rise of a new generation in Germany, the broadcasting of the American miniseries *Holocaust* in 1978, and the 1994 Wehrmacht Exhibition, which sought to dismantle the myth of the "clean" German army during the war — spurred a reevaluation of the Holocaust in Germany and elsewhere. The "guilty few" theory was discredited, and a broader understanding of the role that ordinary Germans — and Poles, Latvians, Lithuanians, Estonians, and others — played in the mass killings slowly came into being. Austria quickly followed its neighbor's lead and abolished its own statute of limitations in the spring of 1965; France had done so for crimes against humanity months before. Three years later, the UN General Assembly adopted Resolution 2391, which stated that crimes against humanity "are among the gravest crimes in international law" and that no statute of limitations should apply in such cases. Fifty-five nations agreed to abide by it.

Five years after the original vote, the German statute was extended for another half decade. Then, on July 3, 1979, the statute came up for a vote in the Bundestag for a third and final time. In attendance that day was the Israeli ambassador to Germany, along with a special guest: Tuviah Friedman sat in the gallery listening to the all-day debate, the culmination of his life's work. Also in the gallery were a group of protestors dressed in concentration camp uniforms; early in the proceedings, the activists stood up and began an impromptu protest, shouting "No freedom for killers!" Police hurriedly ushered them out of the chamber.

Friedman had been campaigning against the amnesty for nearly twenty years. He was fifty-seven now, a bit chubbier, almost completely bald, and still far less famous than Wiesenthal, though that mattered little to him. He hadn't mellowed appreciably in the intervening years; though his stock had risen in Israel after the 1965 vote, he was still irascible and considered something of an obsessive. Now he looked down as the legislators called out their votes, with television cameras pointed directly at the rostrum. Like the first deliberations, these were being carried live on a national TV network. The debate was bitter. One Christian Democrat told his colleagues, "We cannot bow to the ignorance of the American public." (The US House of Representatives had passed a resolution urging further prosecutions of Nazis.) When the last vote was counted, the tally was much closer than in 1965: 255 for the cancellation of the statute, 222 against. But the resistance had won; the statute was abolished forever. "Today . . . ," Friedman wrote, "these Nazi murderers have . . . been silenced until the end of their lives."

In Paris and Tel Aviv, the members of the team read the accounts of the first debate closely, none more so than Mio. Adolf Arndt had drawn a picture of a typical Nazi killer, and it had corresponded closely to Herbert Cukurs; Mio was convinced the mission had deeply affected some members of the Bundestag. "The photograph of Cukurs' corpse, covered in blood and stuck inside a trunk, with the verdict of 'Those Who Will Never Forget' stuck to its chest, was imprinted on the minds of the participants," he said. "I have no doubt that it persuaded some of the floating votes."

The operatives went their separate ways. Zeev Amit, the paratrooper, rejoined the military and was killed in the 1973 Yom Kippur War. Moti Kfir stayed on at Mossad, as did Eliezer Sudit, who participated in many more missions before retiring at age seventy-five. Two years after the mission, Yariv — the brilliant and sociable sabra — was promoted to head of Mossad's European operations.

Mio, the primary actor in the operation, was still the German-born introvert in an agency full of big personalities. After Montevideo, there was newfound respect for his talents, but no promotion. His disappointment was sharp. Mio never headed up his own unit and was given no new responsibilities; when leadership positions came open, he was passed over again and again. "My father wasn't capable of commanding a lot of men," said his son. "But he was frustrated that he wasn't recognized for what he did in the Cukurs mission. They didn't accept him enough for him to go higher." Another Mossad agent called Mio "a general without troops."

He retired in 1981 as the operative who'd worked under more cover identities than anyone in Mossad's history. The news of his role in the mission slowly leaked out. He received letters from the descendants of those killed in the Riga *aktions,* and Israelis would occasionally come up to him on the street and shake his hand, often with tears in their eyes. "People were grateful," said his son. When the team gathered in a Tel Aviv café that spring, a waitress saw them reading a newspaper whose headline announced Cukurs' death. She dropped the tray with their coffee to the floor. She was a survivor from Latvia.

After his retirement, to keep his hand in Mossad, Mio gave lectures to new recruits at the academy. Every so often, he would be called away from his comfortable home in northeast Tel Aviv, driving off through the gate with the steel bar. "If they needed an old fox," said one Mossad agent, "or someone just to stand at the side of an operation and watch, they called him."

In 1997, a *Der Spiegel* journalist tracked Mio down to the palm-lined streets he was fond of strolling along. "He looks like a grandfather from a picture book," the correspondent wrote, "plump, kind and a touch too lenient with the children." Time had leeched away some of the disappointment he felt at being passed over for promotion. What would you tell your descendants if they asked about the most important thing you did in your life? the writer asked. "Without any ado, the truth," he said, "that I killed Herbert Cukurs."

Every year on February 23, Mio and the other team members gathered to mark the operation's anniversary. Yariv often hosted the gathering at his home. His finger, which Cukurs had bitten during the struggle, had fully healed. The mood was always light, and the years continued to produce the harvest of their mission: new trials for old Nazis. The five called the gathering the *yahrzeit,* which means "time of year" in Yiddish and refers to the anniversary of a death; at that time, family members visit the late one's grave and recite the Kaddish. The Mossad men didn't pray for Cukurs, just ate and drank and talked about the events in Montevideo. "My father was very proud of the part he'd played," said Yosef Yariv's daughter, Lihi Yariv-Laor. But they all knew it had been Mio's game. "Of all his missions," his son said, "he was most proud of this one."

Herbert Cukurs was gone, but his essential mystery remained. Why, in the end, did he do it? Even in the memories of Latvian survivors, recorded many years later, there was confusion. In 2000, the Jewish survivor Bernhard Press published a book of his experiences during the Shoah; one anecdote concerned Dr. Weinreich, a Jewish physician who gave Cukurs a clean bill of health before his flight to The Gambia and wished him well. During the war, Press wrote, the two met again in a Riga hospital where the Latvian had come to search for Jews hiding in the beds and closets. Cukurs "forced his way into the women's ward, where he discovered a newborn baby. Births were forbidden in the ghetto. Cukurs snatched the baby by its feet from its mother's bed, smashed its head against the wall so that the skull broke, and tossed the lifeless body to the ground." But immediately after presenting this horrifying image, Press writes, "When the Liepāja ghetto was liquidated, Cukurs spared the lives of Weinreich and Sick, who had once been his own doctors, and sent them to Riga together with a few others. Both of them survived the war." Most Riga Jews believed that it was an implacable hatred that drove Cukurs to his crimes, but it seems his hatred was selective.

To many, Cukurs was bestial, a thing, a berserker. But a few Riga residents registered small protests at the common understanding of him as a garden-variety Jew-hater. "While there is absolutely no doubt that Cukurs was a vicious monster . . . ," said David Silberman, a scholar of the Latvian Shoah, "[he] was a very European type and was likely not a dyed-in-the-wool anti-Semite." Other survivors and their families regarded the question of motivation as disrespectful to the dead. A man who slaughtered Jews for being Jews was by definition an anti-Semite. But some continued to puzzle over the aviator's rapid transformation in the summer of 1941.

The execution in Montevideo seemed to end any hope of a full portrait of the man. After his death, his family, who remained in Brazil, revealed little new information about Cukurs' feelings toward Jews; instead, they told anyone who would listen that he was innocent. Almost nobody believed them, except for a rather large contingent of Latvian nationalists, who lionized the aviator as a patriot and an intrepid explorer. In 2004, members of the right-wing party National Power Unity distributed envelopes bearing Cukurs' portrait; the next year, an exhibit titled *Herbert Cukurs—Presumed Innocent* debuted in Liepāja, his hometown. In 2014, the musical *Cukurs, Herbert Cukurs,* which celebrates the life of the ex-commando, debuted in Latvia. It portrays Cukurs as a "a dashing guy in a chic aviator's uniform," wrote one journalist, "reminiscent of James Bond." Though it features a climactic scene in which Cukurs is surrounded by men and women shouting "Killer!" the play gives the strong impression that his accusers were misguided. When the play reached Riga, rapturous crowds turned out to see it. "What shocked me was that after the show, they all gave it standing ovations," said Boris Mafstir, a Latvian-born Israeli filmmaker whose maternal grandparents were murdered in Riga during the war. "Hundreds of people, they applauded. Suddenly, this Nazi officer can be a national hero."

Cukurs' supporters pointed out that he was never convicted of war crimes in a court of law, while ignoring or dismissing the eyewitness testimonies, including those from other members of the Arājs Commando, detailing his many atrocities. One pro-Cukurs documentary crew even surprised the aging Sasha Semenoff in Las Vegas to question him about his account of the war years and to accuse him of sending an innocent man to his death. In the eyes of his admirers, the aviator was only the latest Latvian victim of the Jews.

And yet, a clear, unimpeachable answer to the deepest enigma about Herbert Cukurs' life did finally emerge years after his execution. And when it did, it came as the result of an oath to the departed.

TWENTY-SEVEN

The Asylum

IN THE LAST YEARS OF THE WAR, Zelma Shepshelovich remained in hiding in the apartment she shared with Nank and the two Arājs killers. She read book after book and took to studying French. In the daytime, before the commandoes gathered in the apartment to drink and tell stories, she could almost believe she was a college student again, learning foreign languages as she'd always dreamt of doing. But her main work in those many months was inscribing in her mind the identities and deeds of the perpetrators of the Latvian Shoah.

After the *aktions*, her situation in Riga remained precarious. Sometime in 1942 or early 1943, Nank, who "hated the Germans" and didn't want to fight for them — and who refused to leave Zelma alone and unprotected — was drafted into the army. If he was called to the Eastern Front, she would be forced out on the street. This distressed him deeply. Hours before his medical exam, he drank cup after cup of coffee and took a handful of speed pills; his heart rate shot up, and the examining doctor declared him unfit to serve. Nank returned to work at the Buildings Department.

Back in their apartment, the two of them remained "constantly frightened." Ieva, Zelma's loyal nanny, was a terrible gossip, and

Zelma worried she would inadvertently reveal Zelma's hiding place. In fact, Ieva had already told a number of her friends and acquaintances — sixteen or seventeen of them in all — that Zelma had survived the *aktions* and was living with her Latvian lover; she'd even mentioned the street address of the apartment.

In February 1943, Zelma was sitting in the kitchen when the doorbell rang. She immediately ran to a small room off the kitchen and hid, then listened as a number of Gestapo officers entered the apartment and began questioning Nank and his roommates, Lidums and Kraujinš, who had been entertaining two Latvian women when the Germans arrived. As the officers checked the women's identity cards, someone opened the door to Zelma's room. She looked up from her chair; it was Nank. "His face was absolutely white," she remembered. "He said, 'Zelma, this is the end.'" He closed the door, and she heard him walk away. Thinking quickly, Zelma snatched up a kerchief like the ones Latvian country girls were fond of wearing, tied it over her hair, slanted her feet on the floor until they were pigeon-toed, picked up a pair of scissors, and began cutting her nails with them. A moment later, one of the Gestapo officers opened the door, saw what appeared to be a servant girl, and asked her who she was. Zelma began speaking in broken German, as if she were a Latvian peasant. The German waved his hand at her in disgust and shut the door. While the Gestapo officers were speaking with Lidums, Zelma dashed out the front door into the frigid air without a coat and managed to hide until the Germans left.

Zelma and Nank were impatient to leave the country. She was eager to alert the world to what Cukurs and the other Latvian commandoes had done and were still doing to the Jews of Latvia. Nank wanted to avoid fighting with the Wehrmacht and also felt he couldn't let Zelma travel alone. He began to save money for their escape. He had recently transferred from the Buildings Department to the railroad administration, and black-market traders regularly paid him 50 deutsche marks apiece for travel permits to carry their illicit goods on Latvian trains. After many months, he'd saved enough to afford two spots on an old fishing boat whose owners had agreed to smuggle a handful of people across the Baltic Sea to Sweden. In April 1944, on the appointed day, the two left their apartment and made their way to Liepāja, Cukurs' hometown, 130 miles to the west of Riga. The sailors who met them there — Latvians who'd deserted the German army — were sopping drunk on vodka. When they saw who their passengers were, they were scornful.

"The man, all right," they told Nank and Zelma. "But we're not going to take a girl with us." The 200-mile crossing was likely to be rough, they said; they didn't want a woman on board.

"Don't worry," Zelma said. "I promise you I will not cry." The sailors finally relented.

After several days of waiting, during which the sailors somehow managed to get even drunker, Nank and Zelma made their escape at night, crawling across the sand under the rotating beam of a lighthouse manned by German soldiers until they reached a small dinghy crowded with twelve other refugees. The sailors pulled at the oars, and the dinghy headed out to sea. They reached the fishing boat and were hoisted aboard.

In the darkness, the moon reflected in the fretted water, their hopes faltered. "There were moments when [the sailors] were so drunk we thought they were going to take us back to Latvia," Zelma remembered. She and Nank huddled together on the deck as the inebriated men steered a zigzag course through the waves. Hours into the journey, Zelma heard the buzz of an engine; as it grew louder, she looked up and saw a German fighter plane above them. In the moonlight, she could make out the pilot's black silhouette. "All of us, including me, were absolutely sure we were going to be shot," she recalled. One of the other refugees dug through Zelma's bag and found a light-blue slip, which might, from a distance, be taken for the Swedish flag. He stood up, waving the slip in the air. The plane turned away without firing on the boat, then disappeared over the horizon.

After sixteen hours on the water, the boat docked in Stockholm, where Zelma, Nank, and the other passengers were welcomed as the first Latvian evacuees to reach the city during the war. The first thing Zelma did in the capital was to seek out a synagogue; in her loneliness, she wanted to see something that had eluded her for months: "a living Jewish person." After one of the first Jews she met listened to her story of survival and her concealment in Nank's apartment, he told her, "Now you have to leave the *goy* and marry a Jewish man." Perhaps he wished to be that man.

"Probably, yes," Zelma replied. "But we are not going to hurry with that." As grateful as she was to Nank, she knew that she couldn't live with him forever. First of all, she wasn't in love with him, and she felt it was her destiny to have a Jewish husband. "To me, [Nank] always was a man who'd saved my life and risked his own," she said. That wasn't enough to build a life on. They were no longer lovers, and eventually, she knew, they would have to part.

Zelma flourished in Stockholm. The city was thick with escaped Jews who were thirsty for adventure and entertainment after years of being hunted. Zelma met many of them. She went to parties with her new friends and took day trips to the countryside. Nank, on the other hand, was unemployed and depressed during their months in Sweden; he barely left their apartment. Their situations had completely reversed; now he was the recluse.

Even as she drank in the beauty of Stockholm, Zelma felt the pull of the promise she'd made. Between studying languages, doing translations for the American embassy, and working in the university library, she sat down and wrote a fifty-five-page report on the deaths of her father and mother, the *aktions,* the mass murders in Riga, Cukurs' wild murder spree — the entire catalogue of terrors of the past three years. It was 1945, and the war was still under way. "The Jewish children and women were being killed," she said. "And I cried it out to all the world." It was the first comprehensive account of the genocide in Latvia and contained a list of the names of all the perpetrators she could remember: Arājs; Cukurs; the men she'd met in Nank's apartment; Lidums and Kraujinš; Arnolds Trucis, the savage killer from 19 Waldemars; and many others. She also went to the refugee camps that had sprung up near Stockholm, looking for the Latvian perpetrators so that she could include their precise locations in the report.

When it was done, Zelma had the document translated into English and made several copies. She called at the American embassy and met with a diplomat there; she handed over the report and explained that Jews were still being hunted in Riga. The diplomat promised to read the report and get back to her. She made her way to the British embassy and then to the offices of the World Jewish Congress and met with officials at both, leaving copies of the document when she departed. After four years of repeating over and over the names, the hometowns, and the life stories of the murderers, she felt unburdened. "I'd kept my word."

In the next few weeks, despite Stockholm's austere charms, she began to grow nervous. There was no call from the embassies, no note in her mailbox on diplomatic letterhead. She suspected that expatriate Latvian diplomats who'd found jobs in the embassies had read the report and reacted badly. "They were terribly frightened," she surmised. "There was a list of a number of war criminals; they destroyed it." Even the World Jewish Congress remained silent. Despite the scrupulous account of names and places and deeds, there was no response to the report. "Nobody in Stockholm, neither the Americans nor the British, believed it."

Zelma decided to go to the newspapers. She condensed the document into a long article, keeping just the most vivid anecdotes, along with the names of the perpetrators, and brought it to *Dagens Nyheter*, one of the leading dailies in the city. The editor in chief invited her into his office and listened to her story. After she'd finished, he spoke. "I am sorry," he said. "For such things, we have no space in our newspaper." The Germans might still win the war, and *Dagens Nyheter* couldn't take the chance of angering them with what was, frankly, a rather incredible story of genocide. There had been a few Holocaust accounts published by then — one of the first in the book *Story of a Secret State,* published in 1944 by a Polish underground fighter — but little or nothing had been revealed about the murders in Latvia.

"I was furious," Zelma said. "I could have killed him for that." She left his office knowing that her mission, and the chance for a full accounting of the crimes of the Latvian Shoah, was over. Years later, back in Riga, she would try again. One day on a streetcar, she spotted the man — she never revealed his name — who'd led the mass killings of Jews in her hometown of Kuldīga. She followed him to the house where he was living, then hurried to the Soviet authorities to file a report. They did nothing.

Soon after the German defeat, Zelma and Nank made plans to return to Latvia, which had been reabsorbed into the USSR. Zelma was determined to find her brother, Zelig, who'd disappeared into the Siberian wastes before the German occupation. She and Nank tried to book train tickets to Riga but were taken instead to a "terrible" refugee camp in Viipuri province, Finland. After weeks in the camp, they made it to Riga, but it quickly became clear to both of them that their lives were in danger. The days of red terror had returned. Roundups, trials, and mass deportations roiled the city. The Kremlin ordered the MGB, the Ministry of State Security, to arrest men and women who'd resisted the 1940 occupation, those who'd spoken out against Soviet rule, "bourgeois nationalists," owners of large farms, suspected subversives, and many others. Latvians who didn't embrace the Soviet takeover were branded "Fascists" and "Nazi collaborators." Zelma grew worried that Nank, who'd made clear his hatred of the Soviets, would be arrested.

A rumor began spreading through Riga: if a Latvian who was being targeted by the MGB married a Jew, he would be saved from the Gulag. It was a continuation of the big lie the Nazis had told, that Jews had helped bring the Soviets to power and wielded excessive influence within the USSR. Now people whispered that anyone protected by a Jew was untouchable. Despite not being in love with Nank, Zelma felt compelled to help the man who'd sheltered her during those dark years. In August 1945, in a simple ceremony, she married him. After four years together, they became man and wife.

The rumor, as it turned out, was false. Being married to a Jew gave one no protection against Soviet terror; it only meant that the Jewish partner was now also suspect. The newlyweds realized they'd made a fatal mistake in returning to Riga. "Immediately the KGB started hunting us," Zelma said, using the name of the agency that succeeded the MGB in 1954. "In those days, if someone came from abroad or knew languages, he was a finished person."

The same month they were married, Nank was arrested by the MGB for distributing anti-Soviet propaganda and taken to prison to await trial. Zelma, who was then unfamiliar with the Soviet theory of justice — "I was stupid and naïve" — went to the public prosecutor and begged for her husband's release. She told the man about how Nank had rescued her, a Jewess, when other Latvians were turning against their neighbors. "I said, 'He is an innocent man and he has saved my life.'" The prosecutor listened impassively. Seeing she was getting nowhere, Zelma changed tack: if they let Nank go, she told the prosecutor, she would fill his spot in the prison. "Take me instead of him," she said.

The prosecutor looked at her with fresh interest. "If you really ask me this," he said in Russian, "we will arrest you." To remain free, Zelma would need to denounce Nank and, preferably, divorce him. She wouldn't hear of it. Zelma went home to brood over Nank's situation, realizing now that "there was no hope he would get out of jail."

There was little left for her in Latvia, but it was almost impossible to leave; the Soviets granted exit papers to very few people. Her parents and sister were dead, her friends murdered or exiled, her brother far away in Russia, where he'd become a wealthy accountant. After returning to Latvia, she'd gone back to her hometown of Kuldīga and found the family's old home occupied by a Latvian family who refused to vacate the house or even discuss payment for it. The mild anti-Semitism that she remembered from her prewar days was now more out in the open, palpable and often vicious. "We understood," she said about herself and Nank, "that we were lost people."

The MGB grew more and more interested in Zelma as the weeks passed; she was harassed, brought in for questioning, detained overnight, accused of being an American, and then an English, spy. She was forbidden to leave the country. "They took me for a very dangerous and very unpredictable person," Zelma said. "Perhaps they were right." Despite the harassment and the growing possibility that she would be charged with political crimes, she refused to leave Nank to face his charges alone. She stayed in Riga.

In late spring, 1946, Nank was brought to a tribunal to receive his sentence. Zelma attended the hearing. The lead judge pronounced a sentence of ten years in the Gulag. As Nank was led out of the chamber, Zelma quietly slipped a gold bracelet off her wrist and handed it to one of the guards. The man brought her to see Nank, who was sitting in a holding cell; they had two or three minutes together. The two embraced, spoke some tender words, and then Nank was taken away. The Soviets had terminated their marriage the moment he was convicted.

Zelma waited for news. Eventually it came: Nank had been taken to Norillag, a labor camp near the city of Noril'sk, above the Arctic Circle. Built on the permafrost, Norillag was among the worst of the labor camps; the sun didn't rise there for three months a year, the inmates breathed in air thick with smoke and ash from the nearby smelters, and temperatures regularly hovered around thirty degrees below zero. One journalist called the place "an unimaginable chamber of horrors, a community of prison camps designed to create nickel and copper industries, and to kill people." When summer arrived each year, the bones of dead inmates would wash out of the melting tundra.

As the months went by, Zelma became convinced that the MGB would arrest her next. She fled to the home of a sympathetic Baptist family, who hid her, but the family had children and Zelma knew she was endangering them by her very presence. She moved on to the home of a Jewish dermatologist and his wife, who managed to obtain false papers for her. After a month, Zelma took the train to her brother's home in a small town called Cherepanovo, close to Kazakhstan; she never spoke about the reception she received from Zelig, but simply having her in his home was an act of courage on his part. There she managed to steal a small bottle labeled POISON that she found in one of the bathrooms. One afternoon, as she was sitting at a table translating a Russian book, the door opened and two MGB colonels walked in. "You are arrested," one of them told her. The charge was "anti-Soviet activities," but it was her link to Nank and her refusal to inform on her friends that were the true offenses. She asked to use the bathroom, taking the bottle of poison with her. After closing the bathroom door, she swallowed its contents. "I'm putting an end to my life," she thought. But the bottle turned out to contain opium, not poison. Zelma was put in the back of a truck beside a heap of coal and taken to an MGB prison in the nearby city of Novosibirsk.

Zelig, whom she had left Sweden to find, now distanced himself from her. "He didn't send me a crust of bread," Zelma later said. After being processed in Novosibirsk, her hair and clothes covered in lice, she was packed onto a freight train back to Riga; she and the other female prisoners were forced to stand upright, their rib cages pressed into each other as they panted for air. Before the cars set off with a clinking sound, the women were fed a dried fish called *vobla;* they began to crave water, but none was offered to them. "Thirst is the worst thing a person can experience," she said. "I saw people go mad."

When they arrived in Riga, she was interrogated again and asked to name friends who had turned against the Soviets. "I counted sixteen people who would go to prison if I told them the names of those who had hidden me. I refused." She was taken to Riga Central Prison, where one day a guard brought her a package containing a spoon made out of cheap metal along with a single egg. Nank had fashioned the spoon in his prison camp, inscribing the number 58 — the article of the Soviet penal code under which he'd been sentenced — and the words ATMINAI NO TAVA NANKA (Remember your Nank). Somehow, he'd learned of her imprisonment and managed to send her these gifts. He also asked his mother and sister to send her food parcels. Zelig, apparently still mortified by her arrest, sent nothing. "Nank was greatly worried about me," she said, "and not about himself."

After many months, Zelma was released; she'd refused to give the MGB the desired names. Perhaps they'd tired of her denials, but she remained under suspicion. Before her arrest, she'd met a Jewish doctor, Grisha Hait, in the Baptist family's home; the two had begun as friends but fell in love over the ensuing months. She wrote Nank to tell him the news. "He knew everything," her daughter Naomi said. Nank was allowed to write one postcard every six months to a relative; he chose to write to Zelma rather than a member of his own family. The postcards continued even after she told him about Grisha. It must have pained him terribly, but Nank would not abandon her over something as insignificant as her marriage to another man.

Zelma and Grisha had a baby girl, Shulamit, a name taken from the Old Testament. Selecting a biblical Jewish name for their child was considered a daring act in Stalin's Russia. Zelma was overjoyed at her daughter's arrival; she breastfed Shulamit and spent hours memorizing every curve and plane of her face. Zelma still dreamt of going to Palestine one day; for the moment, that was out of the question, but her abiding sense of Jewish history — the stories she'd listened to as a child of Jews who'd endured despite living as captives in Babylon and Egypt — stiffened her spine.

Five weeks later, two MGB men came to where she was staying and took her and the child to the Riga insane asylum; her refusal to inform on her friends was used as evidence of her madness. After she and Shulamit spent six months in the asylum, sleeping on a narrow iron cot, MGB agents entered the ward and took the child. They returned Shulamit to Grisha, who was living in "terrible" conditions, unable to find work and nearly destitute because of his connection to his wife. Learning this, Zelma became increasingly distraught. "I decided to commit suicide," she remembered. As she had before the first *aktion,* she collected Ruminol and Veronal tablets; this time, she hid them in the shoulder pads of her dressing gown. One night, she went into the bathroom, taking with her a small photo of Shulamit. She swallowed the pills with a sip of water, then kissed the photo before returning to her cot.

When the inmate in the next bed noticed that Zelma's lips had turned blue, she called for the doctor, who pumped her stomach and saved her life. As punishment, Zelma was sent to the schizophrenic ward. "That was probably the worst time in my life," she said. "Nobody was a human being there." The inmates leapt about the ward "like lions and tigers"; they beat her, clawing at her face ferociously. They saw in her eyes and face someone from their past. Zelma lost weight; she was sick, exhausted, and in a deep depression. Even so, when the MGB agents came to question her, she gave them no names.

After two years, she was released from the asylum and reunited with Grisha and her daughter. Grisha was still unable to work as a doctor and was reduced to laboring in a factory for 70 rubles a month, not enough for the family to live on. Zelma couldn't advertise for work as a teacher of English and German because anyone who spoke those languages was immediately targeted by the Soviet regime. "We were very often hungry," she remembered. Many of their friends, afraid to be associated with a former political prisoner, deserted them. Grisha's relatives, who were well-off and could have helped them financially, cut off contact. The almost absurd string of lucky accidents, sacrifices, and acts of love that had enabled Zelma to survive the German occupation had been broken, and she now experienced the opposite: complete isolation and erasure.

The four of them — Grisha, Zelma, Shulamit, and their Latvian nanny, Ieva, who'd remained faithful to Zelma despite her arrest — lived in a single room in Riga. A second daughter arrived in 1952; Grisha and Zelma called her Naomi, another biblical Jewish name. Zelma never expected to raise her daughters; she believed they would be separated by the MGB before the girls grew much older. "I remember when Naomi was in the cradle, I sang to her because I knew they would take me away again, sitting and crying bitterly because I thought, 'Never, ever will my child know she's Jewish, because she will have no mother.'"

If Zelma had cooperated with the MGB, she would have been able to lead a normal life. But she couldn't abandon her principles, especially when it came to her faith. "I believe in my God," she would tell anyone who listened. One day, when Shulamit was a child, she came home from school wearing the red scarf of the Pioneers, the Soviet youth organization. Joining the Pioneers was a rite of passage, and Shulamit had been excited to sign up. Knowing her mother's hatred of anything associated with the Soviet system, the young girl had brought her elderly Latvian teacher home with her as an ally. When Zelma saw the scarf, she flew into a rage and ripped it off her daughter's neck. "You won't be wearing *this* red rag," she said, before flinging it into the fire. "Mrs. Hait," the teacher said, "you're lucky it's me and not someone else who is here." She never informed the authorities.

The MGB remained interested in Zelma; she expected to be picked up at any moment and taken back to the asylum or to the Gulag. One afternoon, an MGB agent accosted her and brought her to a private house. There another operative offered her a good salary if she would join the MGB and travel back to Sweden, where she still had friends, to work as a spy. "I was terribly upset and terribly nervous. But I said, if you want me to go back to the lunatic asylum, you can send me by force, but I won't do anything for you." She was interrogated for four hours but refused the interrogators' offers and ignored their threats; when she returned home, she was terrified and increasingly depressed. She broke down. "I was always frightened, I was always afraid of people. Here I saw that I couldn't survive, that there was no way to survive." Her only friends were other Jews and artists who, like her, often lived on the margins: opera singers, musicians, painters. "She was hopeless," her daughter said. "My father would say,

'Zelma, don't cry. One day we'll get to Palestine.'" In her lowest moments, Zelma saw no exit. Her only respite came with Stalin's death in 1953, when she was able to resume giving English and German lessons and make a modest living.

In 1956, Nank was released from the prison camp when his ten-year sentence ended. Instead of returning to his relatives in Riga, he went from the train station directly to Zelma's house. Any possibility of romance was long over, but he wanted to be near her, to embrace her, to talk. Nank was frail and often ill; his health had been broken in the Gulag. Grisha took him on as a patient and treated him with what medicines he could find. A few years later, Nank married a Latvian woman who bore a striking resemblance to Zelma. "At first, his wife was terribly jealous of her," said Naomi. "But then they came to our house. They would be at our house every day. And they became great friends." At home, Nank kept a photo of the young Zelma on his piano. Whenever Grisha and Zelma had trouble with the Soviet authorities or just with everyday life, Grisha would say, "We have to ask Nank." He and Zelma shared a kind of love that might be inscrutable to Westerners but that was common under Soviet rule during the purges, a bond formed out of years of exile and bitter suffering.

The war, now more than a decade past, stayed with Zelma. "Not a single night did she sleep," Naomi said. "There were nightmares, hallucinations." Grisha treated her with sedatives, but they rarely worked. In later life, her fellow survivor Sasha Semenoff suffered a different set of symptoms. He found it difficult to trust people he didn't know; the betrayals of Riga had made him suspicious of strangers, especially goyim. But he would slowly page through the book of family photographs he'd carried with him in the Riga ghetto; he'd pause and say a prayer over each loved one. Though he wasn't observant, after a good meal "I would hear him mutter underneath his breath, 'Thank you God for *everything*,'" his son, Paul, said. Zelma hadn't found the same peace. She was, as always, extroverted, dramatic, and intense, and the object of her intensity was often the killers of 1941. At times, she would rave, "I hate them! I hate them! *Let them burn!*" It was as if the memories of the war had turned to acid in her veins.

We don't know when she learned of Cukurs' execution; the Soviet press often publicized stories about Nazi perpetrators, so she may have heard the news soon after it broke. It must have been an electric moment for her: after twenty years, the man who she'd heard confess to three hundred murders had been hunted down, and a just sentence had been carried out in the name of the dead, some of whom were dear to her. The fact that it was Mossad that had achieved this must have thrilled her, too. She'd told the Swedes, the British, the Americans, and the Russians about what Cukurs and the others had done in cold blood. They'd treated her as a silly, fanciful woman. The world had turned her away. Perhaps her testimony had even been among the papers in the file that Mio had studied that day in Paris, the accounts that had spurred the hunt. If so, she would have enjoyed that immensely.

On April 26, 1971, Nank and his wife picked up Zelma and her family and drove them from Riga to the airport in Moscow. It was "the happiest day" she could remember; her family's petition to emigrate to Israel had been accepted (after they had paid a large fee). Latvia for her was a place imbued with blood. Only Nank and Ieva stood apart from those things.

Nank and Zelma understood without saying it that she would never return. Zelma and Grisha embraced him as he wept, then boarded the plane for Tel Aviv.

TWENTY-EIGHT

The Trial

IN TEL AVIV, ZELMA GAVE private language lessons, and Grisha worked as a doctor; they made a new life. Her family flourished; her daughters married and had children themselves. But she still felt, nearly thirty years on, that she'd failed to fulfill her promise to her sister, Paula, and to her father, mother, and the others. She spoke about Arājs and Cukurs and the other perpetrators constantly. "We talked every single day about the war," her daughter Naomi said. "She never came to terms with it." Her children grew up feeling that they, too, had come of age in 1941.

During this time, she received a letter from Laimons Lidums, the Arājs commando with whom Nank had shared the secret of her Jewishness when she hid out in Nank's apartment during the war. "He was looking for an alibi, that he was a nice guy during the German times," Zelma said. "He was scared, and he thought I would be a very good witness in his favor." Lidums was living in Toronto, and Zelma's daughter Naomi was living in Buffalo. When Zelma went to visit her daughter, he invited her to his home. She went.

The ex-commando turned out to be "very friendly," chatting with her about the war years. How eager he was to talk! "I was young and stupid," he said. "This is why I did [those things.]" He took her by the arm and brought her next door to visit his neighbor, a Jewish shop owner. "Here is the lady whose life I saved when she was young!" he announced. She was his Miriam Kaicners. But later, after a few shots of vodka, his memories shifted. He stated categorically that all Jews were Freemasons, and that Stalin and Churchill had been members of the organization as well. The idea that Freemasonry was a front for Jewish world domination has a long history; even Hitler subscribed to parts of the theory, and in 1934 Germany's Federal Ministry of the Interior ordered all Masonic lodges to disband. "Jews are the misery of the world," Lidums told Zelma. Alcohol, as in those days in Riga, was a kind of truth serum.

Zelma said nothing, felt nothing. "I only knew I had a purpose." Nodding at his conspiracy theories, she asked Lidums where his friends, his fellow perpetrators, were living now. What had become of Arnolds Trucis, the man Zelma had seen beating Jews with terrible ferocity? Lidums told her Trucis' address in Philadelphia; she passed the information on to the US Department of Justice, which opened an investigation into the former soldier for war crimes. The Latvian died before his trial could start, but Zelma was satisfied that because of her detective work, he'd experienced a small taste of what her family had gone through in the ghetto. "I thought that this son of a bitch has spent his last ten years in fear. This is what I'm interested in. I don't want them to die peacefully. I want them to die in terror. In danger. It's the only thing I can do."

Zelma finally achieved a small measure of vengeance.

Years passed. In 1976, she spotted a familiar name in an Israeli newspaper. Viktor Arājs had been discovered living in Frankfurt. He was employed as a typographer in a publishing house, going by the name Viktor Zeibots.

Zelma felt her chance had arrived; she went to the Israeli police and told them the story of her parents, about the events of November 30 and December 8, about Cukurs' confession, about 19 Waldemars and the scratches on the wall and the plaque that said CEMETERY OF THE JEWS, and about her rape. The German prosecutors began to build a case based on her testimony and that of a few other Jewish survivors. Three years later, in 1979, Arājs was tried in Hamburg. The German authorities summoned Zelma to testify. She and her younger daughter, Naomi, flew to the city before the trial started. Zelma would be one of the chief witnesses against the former commander. She felt, unusually for her, hesitant and shaky in Hamburg. She'd been given no security—perhaps there was an undercover detective watching over her, but she was never told of it—and spent her nights holed up in a hotel room, repeating dates, names, and incriminating phrases in her mind, then repeating them again. One afternoon when she exited the elevator, she found a drunken old man sitting in the lobby. Could he be a Latvian, watching her and Naomi, waiting for an opportunity to do . . . something? For her, the Jew-hunt had never ended. She and Naomi nervously avoided the man.

There were other Jews in Hamburg that week, waiting in hotel rooms spread across the city and reciting to themselves the details of their own stories. Ella Medalje was among them. After her deliverance from the pits, Latvian friends had provided her with documents that "proved" she was an Aryan, and she'd managed to survive the war. Still, in the intervening years, she'd felt separated from the joys of life by thin glass; nothing seemed quite real to her anymore. Only the Shoah retained its freshness. "My second marriage, the birth of two daughters, all seemed of minor importance compared to such a special story. The only important thing to me in life is the fact that I was able to help administer justice to the horrible butcher." Now, in Hamburg, she was ready to talk about Arājs. In many ways, Zelma felt the same way.

When she entered the courtroom, Zelma spotted the white-haired sixty-six-year-old commando leader sitting at the defense table. "He was acting like a sick man," she remembered, "when he wasn't." Arājs didn't turn to look as Zelma took the stand. Once she was ready, the judge asked Zelma to tell her story, and, sitting erect in her chair, she began to recite the names and dates she'd summoned the night before, the story of the roundups, of Nank and the evenings at his apartment. The Jews in the audience, including a family whose mother had been sterilized during the Shoah and a mysterious Latvian woman whose story Zelma never learned, listened in silence. Zelma told of being brought up from the basement of 19 Waldemars and what followed next.

At the end, the jurists asked her a few questions. Finally, the presiding judge leaned over to address her. "This is hard to believe," he said. "Every word you are saying."

A black-and-white photograph was placed before her, showing about thirty men in uniform.

"Then he was a young man," the judge said. "Now before you an old, ailing man is sitting. Could you identify him?"

Zelma snapped open a case, took out her glasses, put them on. Her eyes ranged down the rows of faces. She pointed at a handsome man, the one with the imperious look. "That's him," she said.

The court was silent. The defense attorney approached her. "How do you know what was done to you was done by Arājs?"

"Because he said to me, 'You bitch, do you know who is standing before you? I am Viktor Arājs.'"

The attorney nodded to the judge, and the cross-examination was over.

Arājs' motivations were never really in doubt. He was a careerist and, at least after mid-1941, a manic anti-Semite; the war had answered two of his deep primary drives. But what about the others, what about Cukurs, on whose story so much history had turned? Later in the trial, Arājs decided to finally reveal, more than thirty years after the war had ended, what had brought the aviator to his little band of killers.

Arājs described their first meeting in Riga. Around the time of the

German invasion, he said, the aviator had left his farm and come to the capital to meet with him. They talked, and Cukurs began "asking for refuge." Rumors were circulating in the countryside, Cukurs told him, rumors claiming that he had worked as a Soviet collaborator during the occupation. The gossip contained details: not only had he turned Bolshevik, but the Soviets had given him a Cadillac as payment for his services.

The aviator was frightened; he must have sensed the murderous rage Latvians felt toward anyone who'd worked with the Russians. He was faced with the same accusation being made wholesale against the Jews in Nazi propaganda being broadcast daily on the radio: *Every Jew is a Bolshevik.* If he didn't find a way to separate himself from the rumors, he could share the Jews' fate. And so he sought out the "famous anti-Semite," Arājs, who took him on as his second-in-command.

Other sources later confirmed the basic facts of his story. In April 1941, during the Soviet occupation, Cukurs had been called to Moscow by Alexander Yakovlev, the designer of a series of Soviet military planes that bore the first three letters of his name in their designations: the Yak-1 single-seater, the shark-nosed Yak-3 frontline fighter, and the Yak-28 bomber. The summons must have come as a thunderbolt for Cukurs. Yakovlev was the undisputed wunderkind of Soviet aviation, a brilliant aeronautical engineer who'd proved just as adept at managing the production of the USSR's airborne fleet as he had been at designing powerful new airplanes. As a young man, like Cukurs, he'd become "sky-crazy"; between 1927 and 1933, Yakovlev had fashioned no fewer than ten new models. By age thirty-four, he'd ascended to near the summit of the Soviet aircraft industry. "Stalin called me for a meeting," Yakovlev remembered. "I was shy but he told me, 'Don't be. We trust you, despite your young age.'"

In 1941, Yakovlev was serving as vice minister of the aviation industry, which gave him the power to put new aircraft designs into production. Stories of Cukurs' exploits and his sharp engineering mind had apparently reached Moscow by this time. The Luftwaffe was far ahead of the Soviets in the crucial area of long-distance bombers; Yakovlev called the flier to Moscow to ask for his help in designing new ones. For Cukurs, it must have revived his dream of transforming the world through his aeronautical genius, which he'd detailed so clearly in his novel.

We can't know Cukurs' mind as he walked out of Yakovlev's office after their meeting, but he was probably balancing a few stark facts. Cooperating with the Soviets, no matter how glorious the prospects might be, meant betraying his homeland. Latvia was suffering terribly under the Soviet occupation; its sons and daughters were being exiled and killed off in the Gulag. But perhaps Cukurs felt he didn't have a choice. Refusing an assignment from an official close to Stalin was, in 1941, already considered unwise. Or perhaps he was blinded by his desire to at last see his pencil-drawn designs rolling off the factory floor and soaring into the air, silver-finned creatures of his imagination spreading east, west, and south. The money and prestige that a collaboration with Yakovlev offered would have been life-changing. Years before, he'd risked his life for fame and glory. He'd enjoyed the acclaim immensely. Here was another chance. He took the assignment.

Months later, with the Wehrmacht at the border and the Red Army in retreat, the whispers began. His daughter, Antinea, remembered the moment Cukurs learned about the rumors circulating in the countryside:

> That day I was playing in the yard at the Lidoņi house and my father was cleaning the pool just in his underpants, when a motorcycle appeared through the lilac, driven by a man with a rifle. There was a short talk with my father. [Soon after,] my father came out of the house dressed in his Latvian army captain's uniform. Both left by the same motorcycle — the man with the rifle in front, father at the back. They returned in a bad mood: someone had told the Germans that father had cooperated too closely with the Russians. What happened

next I don't know. A few days passed. Then father went to Riga to deal with housekeeping affairs.

The aviator must have been terrified. "It did not take much," one historian wrote, "for Arājs' men to suspect one of Communism and take him to the pits." It seems almost certain that the "housekeeping affairs" included the meeting with Arājs, where he asked to join the commandoes. Arājs accepted Cukurs into the unit and even accepted the aviator's offer to donate a car to the outfit. One wonders if it was the Cadillac, which had become a kind of scarlet letter for the Latvian captain.

Arājs' testimony in 1979 finally clarified Cukurs' shift in 1941. It wasn't, after all, a deeply rooted anti-Semitism that drove the former aviator. He betrayed the Jews because if he didn't, he would likely have been murdered alongside them. The sacrifice of those men, women, and children was necessary for him to go on living.

Mio and many others had long believed that the aviator was a member of the anti-Semitic group Pērkonkrusts. That, too, proved to be untrue. Academics have recently begun looking deeply into the organization's records, and they have found no mention of Cukurs in the lists of attendees at Pērkonkrusts meetings in the 1930s. (The name of Viktor Arājs, however, appears several times.) In fact, Cukurs was seen at Riga's prewar cafes, chatting and drinking coffee with Jewish writers. And then there is the curious case of the airplane-shaped chocolates. After his flight to The Gambia, a Latvian company produced candy modeled after his famous aircraft, as a promotion to raise funds for the building of a Jewish-owned sweets factory in Palestine. Why choose Cukurs' plane if he was a well-known anti-Semite?

As it turned out, Cukurs wasn't an outlier among his fellow commandoes. The evidence indicates that very few of them were hardcore Jew-haters. A number of the original members had lost loved ones to the Soviets, and they were eager to have their revenge. Others had joined the unit out of economic desperation; the Soviet occupation had thrown thousands of Latvian men out of work, and the Arājs Commando paid regular wages. A few wanted to play a part in the great unfolding drama. "The war of all wars was thundering all around them," wrote historian Richards Plavnieks. "Bloody revenge was in the air. And a man who wore no uniform was barely a man at all." Plavnieks, the historian who's done the most intensive research into the commandoes' motivations, came away from the archives persuaded that "the Nazis were able to convince a large enough section of the non-Jewish Latvian population that 'Jew' and 'Communist' were interchangeable terms." When members of the Arājs Commando were arrested and interrogated by the Soviets after the war, most of them indicated that it had been anti-Soviet feeling that had driven them to join the group; hatred of Jews was mentioned by only a small fraction. The interviews were conducted during a time when "it was much more dangerous to admit the wish to take revenge on the Soviet regime than to give other reasons," so the men had every reason to attribute the killings to hatred of Jews. But they didn't. In their minds, they were hunting Bolsheviks. Among the 365 members of the Arājs Commando captured by the Soviets, only 4 were found to have been members of the Pērkonkrusts.

Once they were in the unit, the men were subjected to free-flowing torrents of hatred directed mainly at Bolsheviks and their supposed helpers. One member who joined in April 1942 signed up only because he needed work, but he understood what the Arājs Commando stood for. "I knew," he said after the war, "that they were hunting and killing communists and Jews who were devoted to Soviet power."

Even if the men weren't hardened anti-Semites before joining Cukurs and Arājs, there was plainly something at work between the Latvians and the Jews well before the Nazi occupation. Latvians, including members of the Arājs Commando, began helping to round up Jews *only hours after* the Wehrmacht crossed the border and the accusations of Jewish collaboration dominated the airwaves. No propaganda is so effective as to spread so quickly without some receptivity in its audience. Even if most Latvians didn't hate the Jews, everyday emotions—envy, a bit of mistrust—quickly liquefied under the terrific pressure of the Nazi regime into something hot and easily shaped. When one of the commandoes later expressed remorse for what he'd done, he qualified his confession: "Only because of the folly of my youth I took part in the killings . . . The Jews do not need to be exterminated, but should instead be systematically trained to work; in other words, they should be broken of the habit of parasitism so they become useful members of society."

As for Cukurs, the German invasion opened cracks in the aviator's story about himself. He needed a villain to blame, to shroud his own culpability. In fact, *he* was the true villain, the traitor to the country he loved. Who knows exactly what was in the minds of the Holocaust's many executioners? The ancient sediment of anti-Semitism that underlay Latvian culture, as it did so many other nations, exploded with a violence that jumped from the Bolsheviks to the Nazis to collaborators everywhere the Wehrmacht took control. Cukurs was hardly unique; there were many men like him in Poland, Ukraine, Romania, Hungary, and elsewhere. But in his local context, he was the leading monster.

The most convincing portrait of Cukurs as the killing began is not the one that most Jews held, of an authentic, eliminationist Jew-hater, but instead what one memoirist of the war would call a "reptilian opportunist." Perhaps that's what he would have told the five Mossad men who trapped him in Casa Cubertini, had they allowed him to speak. That he never really despised the Jews. That he'd been in a tight spot. That he wished it had all been different. He was still living the great lie of his life until the hammer was raised and the gun fired at the back of his skull.

On December 21, 1979, the judges in the Hamburg court returned their verdict. Viktor Arājs was found guilty of participating in the killing of 13,000 Jews during the second day of the Rumbula massacre and sentenced to life. Arājs spent most of the rest of his life in solitary confinement in a prison in northern Hesse. He died without expressing remorse for his crimes.

Zelma returned to Israel after her testimony. She felt released, at least in part, from the burden she'd carried for three decades. As Ella Medalje put it, "I survived the terrible slaughter and did my holy duty for those who were not destined to live to this day." Zelma continued giving private English lessons and indulged her love for beautiful things: a favorite opera, good clothes, jewelry, and especially flowers, which she kept by her bedside. She had her friends over for long dinners and doted on her students. "To this day, when I happen to meet them, they tell me about the impact she had on their lives," said her daughter Naomi. Despite the torments and nihilism of her first five decades, she felt she'd fulfilled her father's, and her own, deepest wish: to live freely as a Jew in their homeland. "My mother was happy in Israel," Naomi said. "The thought of living among her own people overwhelmed her with positive emotions."

Soon after the trial, Zelma received a call from Laimons Lidums. The Latvian ex-commando was furious; he now realized that Zelma wasn't going to be his alibi for the war years. She might instead help send him to prison. "I will never allow you to tread with your dirty boots on Latvian soil," he said. If she continued to talk publicly about the war and the Latvian atrocities, he would kill her.

Zelma was calm. She had endured too much for such a threat to faze her. "These are not German times when you can threaten me and kill me," she told him. "You are a refugee in Canada. Watch out!" Lidums never called her again.

Throughout the 1970s and 1980s, Zelma wrote letters to Nank, and in return he wrote her "long, affectionate, philososophical letters, contemplating life and fate." His health had never recovered from his years at the labor camp in Noril'sk, and he was unable to work. She sent him and his wife money, food, and clothes whenever she could. Zelma would never return to Latvia, but in 1990 her daughter Naomi went for a visit. It had been twenty years since she'd seen Riga; at her reunion with Nank and his wife, they wept more than they spoke.

Nank grasped her hand and searched her face. "You look like your mother," he said.

On August 19, 1997, Zelma's seventy-seventh birthday, she and Naomi were eating in a café. She was talking fast, gesturing with her hands as she always did, telling a story in her usual dramatic way. She started coughing. A fish bone had gotten caught in her throat. Naomi told her to have some bread, and eventually she got the bread and the bone down and the moment passed.

A telegram arrived the next day telling her that Nank had died the previous afternoon. She and Naomi looked at the time of death and realized he'd passed at almost the exact moment she had felt the bone stick in her throat. An odd, spiky talisman, it spoke to Zelma of the suffering they'd both endured and of their deep connection. "He loved her until his last breath," Naomi said.

ACKNOWLEDGMENTS

My gratitude to those who spoke to me for the book: Naomi Ahimeir, Shulamit Bresler, Paul Semenoff, Gad Shimron, Avner Abraham, George Schwab, Helga Fisch, Lihi Yariv-Laor, Ivar Brod, Stanley Zir, and Zeev Sharon. The staff at Yad Vashem and the United States Holocaust Memorial Museum provided invaluable advice and ferreted out dozens of books, letters, and documents. Martins Zemzaris and Jānis Bošs lent their skills as translators. Richards Plavnieks went out of his way to help me and took the time to read the original manuscript. Paula Oppermann and Guy Walters were equally gracious.

Many thanks to Bruce Nichols, Ivy Givens, and Lisa Glover at Houghton Mifflin Harcourt, and to Mark Robinson for his wonderful jacket design. Barbara Jatkola was the fierce-eyed copyeditor for the book. And thanks, as ever, to Scott Waxman and Ashley Lopez at the Waxman Agency.

NOTES

Prologue: The Apartment on the Avenue de Versailles

page

ix *Mio walked into the lobby*: Anton Kuenzle and Gad Shimron, *The Execution of the Hangman of Riga,* trans. Uriel Masad (London: Vallentine Mitchell, 2004), p. 5. All of the quotes from Mio and Yariv in this chapter and the details about their meeting are from this source. "Anton Kuenzle" was the pseudonym used by Mio, whose given name was Jacob Medad.

x *His son would later say:* Interview with Amnon Meidad.
"I swear to God": Interview with Gad Shimron.

xi *Two other amnesties:* Tuviah Friedman, *The Struggle for the Cancellation of the Statute of Limitations* (Haifa: Institute of Documentation in Israel, 1997), p. 14.

xiv *"Overall, I must say"*: Email interview with David Silberman.

One: The Club on Skolas Street

4 *"The entire Jewish elite gathered"*: Kuenzle and Shimron, *The Execution of the Hangman of Riga,* p. 31.

5 *"As was the custom"*: Ibid.
"Jānis Cukurs and his wife": Ibid.

6 *"would not take off"*: Baiba Šāberte, *Laujiet man runāt!* [Let me

speak!]: *Herberts Cukurs* (Riga: Jumava, 2010), pp. 30–65. Translations by Martins Zemzaris. Unless otherwise noted, all of the quotes and details in this chapter about Cukurs' youth, his journeys to The Gambia and Palestine, and his reception in Latvia on his return are from this source.

10 *"commanded great applause": Osaka Asahi Shimbun,* March 6, 1937.

"Of course, the tiger may eat me": Jaunākās Zinas, January 30, 1937.

13 *"I remember Cukurs speaking":* Kuenzle and Shimron, *The Execution of the Hangman of Riga,* p. 31.

Two: Zelma

15 *It's doubtful that:* USC Shoah Foundation, Visual History Archive, interview with Zelda-Rivka Hait. Her maiden name was Shepshelovich. "Zelma" was her nickname.

16 *bore on his forehead a scar:* USC Shoah Foundation, Visual History Archive, interview with Issack Leo Kram.

"I came home": USC Shoah Foundation, Visual History Archive, interview with Ernest Jacobs.

A Jewish schoolboy once turned: USC Shoah Foundation, Visual History Archive, interview with Edward Anders.

17 *"Of all the territories":* Richards Plavnieks, *Nazi Collaborators on Trial: Viktor Arājs and the Latvian Auxiliary Security Police* (New York: Palgrave Macmillan, 2017), p. 21.

"What they felt in their hearts": USC Shoah Foundation, Visual History Archive, interview with Issack Leo Kram.

"Anything which came from Germany": USC Shoah Foundation, Visual History Archive, interview with Shoshana Kahn.

18 *"is for speaking to the dog":* Ibid.

20 *"Just as intensely and deeply":* Valdis O. Lumans, *Latvia in World War II* (New York: Fordham University Press, 2006), p. 2.

Three: First Night

22 *"He met my father":* United States Holocaust Memorial Museum [hereafter cited as USHMM], oral history interview with Sonja Gottlieb Ludsin, July 13, 1994.

23 "That *put goose bumps on us*": USC Shoah Foundation, Visual History Archive, interview with Shoshana Kahn.
"In the worst case": USC Shoah Foundation, Visual History Archive, interview with Zelda-Rivka Hait.
"It was just a beautiful life": USC Shoah Foundation, Visual History Archive, interview with Sasha Semenoff. Abram Shapiro took the name Sasha Semenoff later in life.

24 *"I will no longer"*: USHMM, oral history interview with George Schwab, March 18, 2005.
"never such innocence again": Philip Larkin, "MCMXIV," first published in *The Whitsun Weddings* (London: Faber and Faber, 1964).
"I am amazed": Bernhard Press, *The Murder of the Jews in Latvia, 1941–1945,* trans. Laimdota Mazzarins (Evanston, IL: Northwestern University Press, 2000), p. 30.
In the town of Jelgava: Ibid., p. 38.

25 *the Russians sent a car:* USC Shoah Foundation, Visual History Archive, interview with Jack Braun.
the wife turned to her husband: USHMM, oral history interview with Julius Drabkin, October 14, 1993.
"This means they have sabotaged it": J. R. Dreyer, *Holocaust Memoir of a Latvian Jew, 1940–45* (Brookline, MA: Self-published, 2011), p. 20.
"I became so terribly distraught": David Silberman, *The Right to Live: A Documentary Account of a Survivor* (New York: Self-published, 2005), p. 41.

26 *"There was no way out"*: USC Shoah Foundation, Visual History Archive, interview with Zelda-Rivka Hait.
"You'd better go home": USC Shoah Foundation, Visual History Archive, interview with Sasha Semenoff.

27 *On the night of June 29:* USC Shoah Foundation, Visual History Archive, interview with Zelda-Rivka Hait.
The men had panicked: Boris Kacel, *From Hell to Redemption: A Memoir of the Holocaust* (Niwot: University Press of Colorado, 1998), p. 2.
"The golden sun": Press, *The Murder of the Jews in Latvia,* p. 44.
One boy ran from his apartment: Kacel, *From Hell to Redemption,* p. 4.

"strong, tanned fellows": Press, *The Murder of the Jews in Latvia*, p. 44.

28 *"It was a strange excitement"*: USHMM, oral history interview with Henry Bermanis, July 12, 1995.

red-and-white Latvian flags: Max Kaufmann, *Churbn Lettland: The Destruction of the Jews of Latvia* (Konstanz, Germany: Hartung-Gorre Verlag, 2010), p. 35.

"I am walking very fast": Dreyer, *Holocaust Memoir*, p. 28.

Latvians stopped pedestrians: Press, *The Murder of the Jews in Latvia*, p. 51.

29 *They were the "inner enemy"*: Andrew Ezergailis, *The Holocaust in Latvia, 1941–1944: The Missing Center* (Washington, DC: United States Holocaust Memorial Museum, 1996), p. 102.

"I awoke on the floor": Ibid., p. 86.

"In the world there is nothing lower": Ibid., p. 159.

"should die as a nation": Ibid., p. 91.

30 *"had nothing to worry about"*: USHMM, oral history interview with George Schwab, March 18, 2005.

"A sea of hatred surrounded us": Press, *The Murder of the Jews in Latvia*, p. 51.

"crooked beaks": Ibid., p. 69.

"What do you want?": USC Shoah Foundation, Visual History Archive, interview with Zelda-Rivka Hait.

31 *"On the first night"*: Silberman, *The Right to Live*, p. 42.

The guards shouted: Kaufmann, *Churbn Lettland*, p. 36.

In the small town of Preili: Ibid., p. 161.

Yiddish-speaking prisoners were taken: There are several mentions of this in testimonies and survivor memoirs, including Press, *The Murder of the Jews in Latvia*, p. 132.

32 *"My mother was like"*: USC Shoah Foundation, Visual History Archive, interview with Sasha Semenoff.

"He liked our flat": Semenoff told the story of Cukurs and the family apartment several times: in ibid. and in his affidavit about Cukurs' crimes, Wiener Library for the Study of the Holocaust & Genocide. This quote is from the affidavit.

33 *"They never saw daylight again"*: USC Shoah Foundation, Visual History Archive, interview with Zelda-Rivka Hait.

Four: Gogol Street

34 *In the town of Bauska:* Kaufmann, *Churbn Lettland,* p. 165.
the Latvians were worker bees: Indeed, Hitler and his lieutenants foresaw that the Latvians, once they'd helped rid their country of Jews, would have to be expelled to make way for the native German settlers who would arrive after the war was won. Lumans, *Latvia in World War II,* p. 147.

35 *"self-purge":* Ezergailis, *The Holocaust in Latvia,* p. 51.
"No obstacles are to be placed": Anton Weiss-Wendt, *On the Margins: On the History of the Jews in Estonia* (Budapest: Central European University Press, 2017), p. 145.
"to take personal care": Andrej Angrick and Peter Klein, *The "Final Solution" in Riga: Exploitation and Annihilation, 1941–1944* (New York: Berghahn Books, 2009), p. 161.

36 *"absolute ruthlessness":* Timothy Snyder, *Bloodlands: Europe Between Hitler and Stalin* (New York: Basic Books, 2012), p. 197.
"They were barefoot": Kaufmann, *Churbn Lettland,* p. 41.
"From every corner of the building": Meir Levenstein, *On the Brink of Nowhere,* trans. Faye Silton (N.p.: Self-published, 1984), p. 4.
"The whole thing was so surrealistic": USC Shoah Foundation, Visual History Archive, interview with Shoshana Kahn.
"People became worse than animals": USHMM, oral history interview with Steven Springfield, March 30, 1990.

37 *On July 4, fifteen or so commandoes:* Cukurs' presence at the burning of the Gogol Street synagogue is contested. He always claimed that he was on his estate until July 14, when he left for Riga. If this date is accurate, he would have been absent for this particular atrocity. Several survivors testified that Cukurs was on Gogol Street that night. Raphael Schub testified in his affidavit about Cukurs' crimes (Wiener Library for the Study of the Holocaust & Genocide) that Cukurs and his men gathered three hundred Latvian Jews in the Great Synagogue, ordering them "to open the ark and spread the Torah scrolls on the synagogue's floor" as they prepared to torch the building. When the Jews refused to follow that order, "Cukurs beat many of them savagely." Others who gave testimonies against Cukurs mentioned the burning of the synagogue. Reuven Salt (file 0.4 152, doc.

204, Yad Vashem) spoke about Cukurs taking part in the burning but did not claim he saw him there with his own eyes. The same is true for Mosher Beilison, who wrote an account for the Brazilian periodical *The Journal* describing the events of that night (file 0.4 152, doc. 89, Yad Vashem).

As for Schub, his testimony in another Holocaust case, the 1950 prosecution of the man known as Fritz Scherwitz, commandant of the Lenta work camp in Latvia, was called into question and eventually found not to be credible. Schub "attributed deeds and motives to Scherwitz that were incongruous and patently untrue, all of them hearsay, for Schub had never set foot in Lenta." Gertrude Schneider, ed., *The Unfinished Road: Jewish Survivors of Latvia Look Back* (Westport, CT: Greenwood Publishing, 1991), p. 75.

But Abram Shapiro, in his oral history, placed Cukurs in Riga on July 2, one of the first nights after the German occupation. Since Cukurs lied consistently about every aspect of his behavior during the war, he is hardly a credible source on this point. USC Shoah Foundation, Visual History Archive, interview with Sasha Semenoff.

"Since the people of Riga": Rudīte Vīksne, "Members of the Arājs Commando in Soviet Court Files: Social Position, Education, Reasons for Volunteering, Penalty," in *The Hidden and Forbidden History of Latvia Under Soviet and Nazi Occupation* (Riga: Institute of the History of Latvia, 2005), p. 201.

38 *"Devout Jews from the vicinity"*: Angrick and Klein, *The "Final Solution" in Riga*, p. 72.

"Life is indescribable": Dreyer, *Holocaust Memoir*, p. 37.

Jews developed stutters: USHMM, oral history interview with Henry Bermanis, July 12, 1995.

Some Jewish girls: Frida Michelson. *I Survived Rumbuli* (New York: Holocaust Library, 1979), p. 36. Michelson uses the alternate spelling "Rumbuli" (instead of "Rumbula") in her book.

Arkadi Schwab, the doctor: Interview with George Schwab.

"As we went to work": Kaufmann, *Churbn Lettland*, p. 84.

39 *"harvest some Jews"*: USHMM, oral history interview with Edward Anders, February 28, 1997.

after Cukurs arrested Pinchas Shapiro: USC Shoah Foundation, Visual History Archive, interview with Sasha Semenoff. All of the quotes and details in this chapter about Abram's experiences during the German occupation are from this source.

41 *"would have been afraid"*: USHMM, oral history interview with Julius Drubkin, April 9, 1992.

"Completely impoverished": Angrick and Klein, *The "Final Solution" in Riga,* p. 102.

Zelma was now working: USC Shoah Foundation, Visual History Archive, interview with Zelda-Rivka Hait. All of the quotes and details in this chapter about Zelma's experiences during the German occupation are from this source.

43 *"A new world!"*: Kaufmann, *Churbn Lettland,* p. 81.

the light of a single match: Levenstein, *On the Brink of Nowhere,* p. 27.

Even non-Latvian Jews: Angrick and Klein, *The "Final Solution" in Riga,* p. 118.

44 *packed into the stairways:* Levenstein, *On the Brink of Nowhere,* p. 33.

When one child: Max Kaufmann, "The War Years in Latvia Revisited," in *The Jews in Latvia,* ed. M. Bobe (Tel Aviv: Association of Latvian and Estonian Jews in Israel, 1971), pp. 351–68.

"He was drunk": Silberman, *The Right to Live,* pp. 51–52.

"He told her to open her mouth": Testimony of Reuven Barkan, Viktor Arājs trial, Hamburg State Archive.

45 *"shouted that they needed"*: Testimony of Hanna Jakobson, Viktor Arājs trial, Hamburg State Archive.

"These girls never came back": Ibid.

"People didn't want": Kaufmann, *Churbn Lettland,* p. 111.

"In my wildest dreams": Kacel, *From Hell to Redemption,* p. 6.

"The greatest tragedy": Benjamin Lieberman, *Terrible Fate: Ethnic Cleansing in the Making of Modern Europe* (Chicago: Ivan R. Dee, 2006), p. 185.

46 *"I did not want to believe"*: Kacel, *From Hell to Redemption,* p. 35.

Five: 19 Waldemars

47 *Max Tukacier, who'd known Cukurs:* Affidavit of Max Tukacier, Wiener Library for the Study of the Holocaust & Genocide. All of the

quotes and details in this chapter about Tukacier's experiences during the German occupation are from this source.

CEMETERY OF THE JEWS: Testimony of Gari Rotov, mechanic in the garage of 19 Waldemars Street, Latvian State Archive.

50 *One night, Ella Medalje:* Testimony of Ella Medalje, Viktor Arājs trial, Hamburg State Archive. Unless otherwise noted, all of the quotes and details in this chapter about Medalje's experiences during the German occupation are from this source.

Cukurs' men had ripped off: Testimony of Abram/Abraham Lipchin to the Israeli Untersuchungsstelle, September 1, 1978, Hamburg State Archive.

51 *"Jewish girls, come out!":* USC Shoah Foundation, Visual History Archive, interview with Zelda-Rivka Hait. All of the quotes and details in this chapter about Zelma's experiences during the German occupation are from this source.

55 *"the shovels . . . smeared with blood":* Testimony of Sasha Semenoff, Viktor Arājs trial, Hamburg State Archive.

Six: The Moscow Suburb

56 *"peeling potatoes in the kitchen":* USC Shoah Foundation, Visual History Archive, interview with Zelda-Rivka Hait. All of the quotes and details in this chapter about Zelma's experiences in the Riga ghetto are from this source.

58 *a much smaller number:* Kaufmann, "The War Years in Latvia Revisited."

"You were saved the first time": USC Shoah Foundation, Visual History Archive, interview with Sasha Semenoff.

"The ghetto will be liquidated": Kaufmann, *Churbn Lettland,* p. 59.

59 *"The decree hit the ghetto":* Ibid.

61 *"ringing, crackling frost":* Michelson, *I Survived Rumbuli,* p. 76.

"in motion like ants": Ezergailis, *The Holocaust in Latvia,* p. 247.

62 *they shed their dresses:* USC Shoah Foundation, Visual History Archive, interview with Yakob Basner.

the opposite conclusion: Ezergailis, *The Holocaust in Latvia,* p. 247.

"Who will avenge us": Press, *The Murder of the Jews in Latvia,* p. 102.

Abram Shapiro was in the small ghetto: USC Shoah Foundation, Visual History Archive, interview with Sasha Semenoff.

Seven: November 30

64 *"You have thirty minutes":* Ezergailis, *The Holocaust in Latvia,* p. 249.
"rolling drunk": Press, *The Murder of the Jews in Latvia,* p. 103.

65 *"Children were hidden":* Levenstein. *On the Brink of Nowhere,* p. 28.
Only the occasional cry: Press, *The Murder of the Jews in Latvia,* p. 105.

66 *As the Jews streamed by:* Kaufmann, *Churbn Lettland,* p. 125.
"We are not going": USHMM, oral history interview with Carolina Knoch Taitz, November 13, 1990.
cords of wood: Ibid.

67 *"Already operated!":* Angrick and Klein, *The "Final Solution" in Riga,* p. 141.
"shot empty": Ibid.
On a road near: Ezergailis, *The Holocaust in Latvia,* p. 251.
"A Jewish woman started screaming": Affidavit of Issack Leo Kram, Wiener Library for the Study of the Holocaust & Genocide.
When they finished: Testimony of Aron Barinbaum, Central Archive of the Russian Federation.

68 *As the men tramped back:* Testimony of Aron Prejl, Latvian State Archive.
"I remember it was squashing": USHMM, oral history interview with Carolina Knoch Taitz, November 13, 1990.

69 *"It was like a miracle":* USC Shoah Foundation, Visual History Archive, interview with Sasha Semenoff.
"He knew that": USC Shoah Foundation, Visual History Archive, interview with Zelda-Rivka Hait. All of the quotes and details in this chapter about Zelma's experiences in the Riga ghetto are from this source.

70 *"denounced him as a Jew":* USHMM, oral history interview with George Schwab, March 18, 2005.
"You can feel it": USC Shoah Foundation, Visual History Archive, interview with Vincent Benson.

Eight: The Valley of the Dead

74 *"Papa, I'm afraid"*: USC Shoah Foundation, Visual History Archive, interview with Shoshana Kahn.

"Where will you bring us?": Ezergailis, *The Holocaust in Latvia,* p. 258.

75 *smashing their skulls:* It should be noted that these stories remain controversial and may have been exaggerated by some survivors. Though multiple eyewitnesses saw Cukurs shoot young children to death, some Latvians and scholars have disputed the testimony of his smashing skulls in this way. Despite these reservations, the anecdotes became emblematic among the Jewish survivors of Cukurs' behavior during the *aktions* and were repeated far and wide. But, of all the stories about the aviator, they are the ones whose credibility has been most often questioned.

perhaps over one thousand bodies: Angrick and Klein, *The "Final Solution" in Riga,* p. 155.

"The people seemed indifferent": Silberman, *The Right to Live,* p. 59. All of the quotes and details about Medalje's experiences at the pits are from this source.

76 *a naked woman stopped:* Angrick and Klein, *The "Final Solution" in Riga,* p. 156.

"Drop all your valuables": Michelson, *I Survived Rumbuli,* p. 89. All of the quotes and details about Michelson's experiences at the pits are from this source.

77 *A construction specialist:* Angrick and Klein, *The "Final Solution" in Riga,* p. 133.

"still writhing and heaving": Ezergailis, *The Holocaust in Latvia,* p. 253.

They were ordered to hold: Ibid., p. 294.

80 *"It's happening again"*: USC Shoah Foundation, Visual History Archive, interview with Sasha Semenoff.

81 *The next morning:* Ezergailis, *The Holocaust in Latvia,* p. 256.

82 *The trucks vanished:* USC Shoah Foundation, Visual History Archive, interview with Zelda-Rivka Hait. All of the quotes and details in this chapter about Zelma's experiences after the *aktions* are from this source.

Nine: A Latvian in Rio

87 *In early 1942:* Testimony of Genadijs Murnieks, Latvian State Archive.
They fought against Stalin's battalions: Andrew Ezergailis, "Sonderkommando Arājs" (paper presented at the Ninth International Conference on Baltic Studies in Scandinavia, Stockholm, June 3–4, 1987), p. 20.
"The streaks of luminous bullets": Laikmets, no. 39 (September 25, 1942), accessed via Latvijas Nacionālā Digitālā Bibliotēka.
Cukurs departed the Arājs Commando: Testimony of Eduard Schmidts, 1948, Central Archive of the Russian Federation; testimony of Arnis Upmalis, former member of the Arājs Commando, 1975, Latvian State Archive.

88 *he enlisted in the Latvian Legion:* Linng Cardozo and Marcelo Silva, *El baúl de Yahvé: El Mossad y la ejecución de Herberts Cukurs en Uruguay* (Montevideo: Carlos Alvarez, 2012), p. 53.
he returned to Latvia: Testimony of Robert Purinsh, former member of the Arājs Commando, 1948, Central Archive of the Russian Federation. Purinsh said that he saw Cukurs "in Kurzeme" in 1943. "Kurzeme" is the Latvian name for Courland, the former duchy of Duke Jacob of Courland, and refers to the westernmost region of Latvia, which includes Cukurs' hometown of Liepāja. As for his stay in Germany, multiple sources, including Cardozo and Silva, *El baúl de Yahvé,* place him there before the war's end.
The family made their way: Šāberte, *Laujiet man runāt!,* pp. 111–15. Unless otherwise noted, all of the quotes and details in this chapter about Cukurs' experiences in the latter half of the war, his postwar movements, and his early days in Brazil are from this source.

89 *"The Nazis discovered her":* Michael Bar-Zohar, *The Avengers: The Drama of the Daring Jews Who Are Avenging the Six Million Dead* (New York: Hawthorn Books, 1967), p. 262.

90 *"They showed great respect":* Bar-Zohar, *The Avengers,* p. 262.

91 *"The SS Obersturmführer Cukurs":* Statement from the Public Relations Sub-Committee of the Association of Baltic Jews in Great Britain, Paris, May 23, 1946, Yad Vashem, file 0.4 152.

ACCORDING STATEMENT: Radiograma telegram to Hoff Brasfur, São Paulo, March 27, 1950, Yad Vashem, file 0.4 152.
"There is no doubt": Letter from Baron F. Elwyn-Jones, MP, to Mr. E. Michelson, Committee for the Investigation of Nazi Crimes in Baltic Countries, May 18, 1949, app. C, Herbert Cukurs case file, Wiener Library for the Study of the Holocaust & Genocide.

92 *imprisoned in the Lenta concentration camp:* Ed Koch, "For Sasha Semenoff, Holocaust Survivor and Longtime Vegas Performer, 'Music Was His Life,'" *Las Vegas Sun,* January 11, 2013.

93 *In early 1949:* Kuenzle and Shimron, *The Execution of the Hangman of Riga,* p. 46.

94 *A Jew known only as "Victor":* Letter to Herr Michelson, Stockholm, January 11, 1949, Herbert Cukurs case file, Wiener Library for the Study of the Holocaust & Genocide.

Ten: "The Epitome of Humanity"

95 *"in all social sectors":* Unless otherwise noted, this chapter is based on the testimony of Milda Cukurs, in Cardozo and Silva, *El baúl de Yahvé,* chap. 7; interview with Helga Fisch; and "Report on the Cukurs Case," Records of the World Jewish Congress, USHMM.

97 *"I am not quite happy":* Guy Walters, *Hunting Evil: The Nazi War Criminals Who Escaped and the Quest to Bring Them to Justice* (New York: Broadway Books, 2010), p. 321.

98 *"The matter of Cukurs":* Letter from Max Kaufmann to unknown recipient in Brazil, n.d., file 0.4 152, Yad Vashem.
To get to the bottom: Statement of Miriam Kaicners, file 0.4 152, doc. 69, Yad Vashem. The members of the commission also included details of their interview in a letter to Max Kaufmann, which is included in the file.

99 "Cukurs der Judenmorder": The epithet is from a letter to the Ambassador of Brazil to the United States, August 22, 1950, Records of the World Jewish Congress, USHMM. German-speaking Jews often used the phrase to refer to Cukurs.
Cukurs himself disputed: Undated newspaper article, file 0.4 152, doc. 27, Yad Vashem. Cukurs did give one interview to the Brazilian journalist Eduardo Ramalho, in which he supposedly confessed, saying, "I cannot deny that I killed several Jews." But it's likely he was

talking about his time with the Arājs Commando in western Russia, where he was fighting Russian partisans, some of whom were Jewish. Cukurs steadfastly denied killing Jews in Latvia and outside the theater of war. The reference to the Ramalho interview can be found in *Nossa Via,* August 18, 1960, file 0.4 152, doc. 139, Yad Vashem.

100 *"When speaking of Cukurs":* Email interview with Paul Semenoff.

In March 1941: Brīvais Zemnieks, no. 64 (March 14, 1941), accessed via Latvijas Nacionālā Digitālā Bibliotēka.

101 *"Cukurs . . . cannot disguise":* Undated document, file 0.4 152, doc. 81, Yad Vashem. The psychologist's name was Eliezer Schneider.

"I'm . . . drowning": Laiks, no. 50 (October 25, 1950), accessed via Latvijas Nacionālā Digitālā Bibliotēka.

"When I heard": Legendu mednieki [Legend hunters], directed by Eduards Majauškis (2008), documentary.

"Cukurs gathered his children": Cukurs' eldest son, Ilgvars, was apparently not living in Brazil with the family at the time.

"Even if I was a soldier": Ibid.

102 *"No one who flies": Folha de S. Paulo,* August 6, 2006.

"Why is he walking": Šāberte, *Laujiet man runāt!,* p. 115.

For Antinea, it was a sign: There is a potential problem with Antinea's story. Jews in Latvia were not typically issued striped prisoners' clothing before the Kaiserwald concentration camp opened north of Riga in the spring of 1943. It's unclear why a Jew would have been dressed in prisoners' clothing a full year before that date.

"his life was endangered": "Report on the Cukurs Case."

103 *"From Riga, we send a warning": Nossa Voz,* July 29, 1963.

Eleven: Anton Kuenzle

104 *"He really felt":* Interview with Gad Shimron.

"I remembered everything": Kuenzle and Shimron, *The Execution of the Hangman of Riga,* p. 4. All of the quotes and details about Mio's life in this chapter are from this source.

105 *Mossad told his family:* Interview with Amnon Meidad.

Mossad agents talk: Interview with Gad Shimron.

106 *"He's really going":* Ibid.

"Can you imagine": Gad Shimron, in *Nazi Fugitives,* season 1, epi-

sode 1, "Herbert Cukurs," directed by Tim Wolochatiuk, American Heroes Channel, 2010.

"Everyone had great hopes": The story of the agent in Beirut is from Stuart Steven, *The Spy-Masters of Israel: The Definitive Account of the Intelligence Chiefs Who Helped Shape the Destiny of a Nation* (New York: Scribner, 1981), p. 164.

107 *"If when you're here"*: Anne Barker, "Mossad 'Factory' Churned Out Fake Australian Passports," Australian Broadcasting Corporation, February 25, 2010, https://www.abc.net.au/news/2010-02-26/mossad-factory-churned-out-fake-australian/343612.

a fully stocked "travel department": Gordon Thomas, *Gideon's Spies: The Secret History of the Mossad* (New York: St. Martin's, 2015), p. 665.

"every kind of ink": Barker, "Mossad 'Factory' Churned Out Fake Australian Passports."

110 *When one agent:* Interview with Zeev Sharon.

111 *"For Israel, he could give"*: Interview with Amnon Meidad.

Twelve: The Merciless One

112 *"Listen, you bastard!"*: Tuviah Friedman, *The Hunter: The Autobiography of the Man Who Spent Fifteen Years Searching for Adolf Eichmann* (Whitefish, MT: Kessenger, 2010). Unless otherwise noted, all of the quotes from Friedman in this chapter and the details about his youth and experiences during World War II are from this source.

114 *he even hunted down:* Friedman obituary, *The Telegraph* (UK), February 16, 2011.

115 *"My beloved Mama!"*: For a selection of the notes, see "Last Notes on the Walls of the Great Synagogue," JewishGen KehilaLinks, https://kehilalinks.jewishgen.org/kovel/notesonwall.htm.

"One so wants to live": Snyder, *Bloodlands,* p. 223.

117 *"He pestered and annoyed"*: Review of Friedman's *The Hunter, New York Times,* March 19, 1961.

118 *"I had come to the conclusion"*: Friedman, *The Struggle for the Cancellation of the Statute of Limitations,* p. 2. The subsequent quotes from Friedman in this chapter and the details about his campaign against the statute are from this source.

Thirteen: The Late One

122 *In Paris, Mio went again:* Unless otherwise noted, this chapter is based on Kuenzle and Shimron, *The Execution of the Hangman of Riga,* pp. 20–24, 48–54; Abraham Rabinovich, "Executing the Hangman," *Jerusalem Post,* March 5, 2010; and "Dies ist Mein Mörder," *Der Spiegel,* July 28, 1997.

123 *In Damascus:* The details about Cohen and his mission are from Zwy Aldouby and Jerrold Ballinger, *The Shattered Silence: The Eli Cohen Affair* (New York: Coward, McCann & Geoghegan, 1971).

Fourteen: First Contact

127 *Mio returned to the marina:* Unless otherwise noted, this chapter is based on Kuenzle and Shimron, *The Execution of the Hangman of Riga,* pp. 55–61; Rabinovich, "Executing the Hangman"; and "Dies ist Mein Mörder."

129 *Mio waved this away:* Testimony of Gūnars Cukurs, in Cardozo and Silva, *El baúl de Yahvé,* chap. 7.

Fifteen: The Campaign

135 *"This process":* Simon Wiesenthal and Ewald Osers, *Justice Not Vengeance: Recollections* (New York: Grove Weidenfeld, 1989), p. 160.

136 *"I realized," said Wiesenthal:* Ibid., p. 161.

"an unprecedented injustice": Hella Pick, *Simon Wiesenthal: A Life in Search of Justice* (Boston: Northeastern University Press, 1996), p. 208.

"The Jew Pig, Austria": Tom Segev, *Simon Wiesenthal: The Life and Legends* (New York Doubleday, 2010), p. 8.

137 *If Wiesenthal didn't stop:* Pick, *Simon Wiesenthal,* p. 160.

"had scattered": Ibid.

"Prosecutors had filed": "Historian Exposes Germany's Minute Number of Convictions for Nazi War Crimes," *Times of Israel,* November 10, 2018.

"the dumbest Nazis": Ibid., p. 111.

138 *The responses came:* Ibid., p. 191.

MORAL DUTIES HAVE NO TERM: Ibid., p. 208.

Mio flew to Brasília: The rest of this chapter is based on Kuenzle and Shimron, *The Execution of the Hangman of Riga,* pp. 62–67.

Sixteen: "Our Own Thomas Edison"

140 *Back in Paris:* Unless otherwise noted, this chapter is based on Kuenzle and Shimron, *The Execution of the Hangman of Riga,* pp. 69–75; Rabinovich, "Executing the Hangman"; and "Dies ist Mein Mörder."

"If you passed him": Interview with Zeev Sharon.

141 *The bomb exploded:* Yosef Evron, "Remembering Eliezer Sudit," a privately published remembrance acquired from a relative of Sudit.

143 *"He was holding":* Herbert Cukurs, *Starp zemi un sauli* [Between the earth and the sun] (Riga: Valters un Rapa, 1937), unpaginated. Translations by Jānis Bošs.

Seventeen: The Plantation

145 *Mio left his hotel:* Unless otherwise noted, this chapter is based on Kuenzle and Shimron, *The Execution of the Hangman of Riga,* pp. 76–85; Rabinovich, "Executing the Hangman"; "Dies ist Mein Mörder"; and Cardozo and Silva, *El baúl de Yahvé,* pp. 123–37.

151 *"Mossad had never done":* Interview with Gad Shimron.

"He saw the mission": Interview with Amnon Meidad.

Eighteen: Paranoia

156 *"excessive agitation": Jerusalem Post,* February 19, 1965.

"German embassies and consulates": Los Angeles Times, February 7, 1965.

In the raw cold: New York Times, January 15, 1965.

The activist even secured: C. David Heymann, *Bobby and Jackie: A Love Story* (New York: Atria, 2009), p. 123.

157 *"What on earth":* Friedman, *The Struggle for the Cancellation of the Statute of Limitations,* p. 12.

"Because this seemed": Segev, *Simon Wiesenthal,* p. 189.

158 *"Approximately ninety percent":* Frank Buscher, "I Know I Also Share the Guilt: A Retrospective on the West German Parliament's 1965 Debate on the Statute of Limitations for Murder," *Yad Vashem Studies* 34 (2006): 257.

159 *Back at his hotel:* Unless otherwise noted, the rest of this chapter is based on Kuenzle and Shimron, *The Execution of the Hangman of Riga,* pp. 86–92.

161 *"a product of":* Alfons Klein and Earl W. Kintner, *Trial of Alfons Klein, Adolf Wahlmann, Heinrich Ruoff, Karl Willig, Adolf Merkle, Irmgard Huber, and Philipp Blum* (London: William Hodge, 1949), p. 170.

163 *"He was a fat guy":* Testimony of Gūnars Cukurs, in Cardozo and Silva, *El baúl de Yahvé,* chap. 7.
"He always talked": Testimony of Milda Cukurs, in ibid.

165 *"It happened during the lunch":* Ibid.

Nineteen: The Sabras

167 *The summer tans:* Unless otherwise noted, this chapter is based on Kuenzle and Shimron, *The Execution of the Hangman of Riga,* pp. 99–107; Rabinovich, "Executing the Hangman"; and "Dies ist Mein Mörder."

169 *"He was a very brave guy":* Interview with Zeev Sharon. All of the quotes from Zeev Sharon in this chapter are from this source.

173 *"It's not a movie":* Ibid.

Twenty: "Certain Categories of Murder"

174 *Imi Lichtenfeld drove the men:* Unless otherwise noted, this chapter is based on Kuenzle and Shimron, *The Execution of the Hangman of Riga,* pp. 99–107; Rabinovich, "Executing the Hangman"; and "Dies ist Mein Mörder."

177 *Just after 7 a.m.:* Aldouby and Ballinger, *The Shattered Silence,* p. 9.

178 *The signs varied: Jerusalem Post,* March 7, 1965.
"subdued sobbing": The Guardian (UK), March 1, 1965.
"I am the only survivor": Toronto Globe and Mail, February 17, 1965.

179 *"Their barbarous acts": New York Times,* February 24, 1965.
Simon Wiesenthal received: Wiesenthal and Osers, *Justice Not Vengeance,* p. 160.
"freeing thousands of anti-Jewish patriots": Toronto Globe and Mail, January 29, 1965.
"is hereby sentenced to death": "Threat to Bonn Parliament Traced to Neo-Nazi Network," *New York Times,* May 28, 1965.

"certain categories of murder": Buscher, "I Know I Also Share the Guilt," p. 272.
"dominated and indeed blackmailed": Ibid., p. 278.
180 *57 percent of Germans supported the amnesty:* "Fight Looms at Bonn on Nazi Trial Bills," *Washington Post,* March 20, 1965.
"ugly, elderly, schoolmasterish man": *Jerusalem Post,* March 16, 1965.
BUCHER STAKES JOB ON STATUTE: *Jerusalem Post,* January 25, 1965.
"most vociferous opponent": *The Guardian* (UK), February 1, 1965.

Twenty-One: The Camera

182 *On January 28:* Unless otherwise noted, this chapter is based on Kuenzle and Shimron, *The Execution of the Hangman of Riga,* pp. 108–13; Rabinovich, "Executing the Hangman"; and "Dies ist Mein Mörder."
185 *"If something happens"*: "Dies ist Mein Mörder."
"If someone comes after me": *Legendu mednieki* [Legend hunters].
187 *On February 15:* Michael Bar-Zohar and Nissim Mishal, *Mossad: The Greatest Missions of the Israeli Secret Service* (New York: Ecco, 2012), p. 182.

Twenty-Two: "To Live With a Few Murderers"

188 *"almost unparalleled"*: Buscher, "I Know I Also Share the Guilt," p. 277.
"in individual cases": Ibid., p. 278.
189 *"We must be prepared"*: Caroline Sharples, "In Pursuit of Justice: Debating the Statute of Limitations for Nazi War Crimes in Britain and West Germany During the 1960s," *Holocaust Studies* 20, no. 3 (2014): 81–108.
"a shocking reminder": Hannfried von Hindenburg, *Demonstrating Reconciliation: State and Society in West German Foreign Policy Toward Israel, 1952–1965* (New York: Berghahn Books, 2007), p. 82.
"sacrilege to millions": Sharples, "In Pursuit of Justice."
"had to fight": Hindenburg, *Demonstrating Reconciliation,* p. 83.
190 *"it is very unlikely"*: Ibid., p. 81.
The four Mossad team members: Unless otherwise noted, the rest of this chapter is based on Kuenzle and Shimron, *The Execution of the Hangman of Riga,* pp. 114–20; Rabinovich, "Executing the Hangman";

"Dies ist Mein Mörder"; Cardozo and Silva, *El baúl de Yahvé*, pp. 123–37; and interviews with Zeev Sharon and Amnon Meidad.

196 *"We were impelled"*: "Hitler to Be 'Prosecuted,'" *Irish Times*, February 20, 1965.

"if Adolf Hitler returns": *Atlanta Constitution*, February 20, 1965.

HITLER FACES COURT: *Washington Post*, February 20, 1965.

MACABRE MOVE: *South China Morning Post*, February 20, 1965.

198 *"I was always a brave man"*: Bar-Zohar and Mishal, *Mossad*, p. 182.

Twenty-Three: The House on Colombia Street

199 *Cukurs was scheduled to arrive:* Unless otherwise noted, this chapter is based on Kuenzle and Shimron, *The Execution of the Hangman of Riga*, pp. 121–28; Rabinovich, "Executing the Hangman"; "Dies ist Mein Mörder"; Cardozo and Silva, *El baúl de Yahvé*, pp. 123–37; and interviews with Zeev Sharon, Gad Shimron, and Amnon Meidad.

206 *It was an eyewitness account:* Roderick Stackelberg and Sally A. Winkle, eds., *The Nazi Germany Sourcebook: An Anthology of Texts* (Abingdon-on-Thames: Routledge, 2013), p. 358.

"Without screaming or weeping": From Trial of the Major War Criminals before the International Military Tribunal, Volume XVIX, published in accordance with the direction of the International Military Tribunal by the Secretariat of the Tribunal, under the jurisdiction of the Allied Control Authority for Germany. Digital document retrieved from the Library of Congress. LOC Call Number KZ1176.T748 1947.

Twenty-Four: The Wait

209 *In the black Beetle:* Unless otherwise noted, this chapter is based on Kuenzle and Shimron, *The Execution of the Hangman of Riga*, pp. 129–39; Rabinovich, "Executing the Hangman"; and "Dies ist Mein Mörder."

212 *"Mio hated journalists."*: Interview with Amnon Meidad.

Twenty-Five: An Offer

215 *The story flashed:* Unless otherwise noted, this chapter is based on Kuenzle and Shimron, *The Execution of the Hangman of Riga*, pp. 129–39; and Cardozo and Silva, *El baúl de Yahvé*, pp. 193–210.

"Scattered throughout Latin America": Jack Anderson, "The End of a Nazi," *St. Louis Dispatch,* May 2, 1965.

"The death of Herbert Cukurs": Ibid.

In Montevideo: Jewish Telegraphic Agency, March 10, 1965.

216 *"Frank Sinatra's favorite violinist"*: Koch, "For Sasha Semenoff."

"I would hear Cukurs laughing": *Nazi Fugitives,* season 1, episode 1, "Herbert Cukurs," directed by Tim Wolochatiuk, American Heroes Channel, 2010.

"They got him!": Interview with Paul Semenoff.

218 *"We must not let"*: *Chicago Tribune,* May 10, 1965.

"the perpetrator has become": Buscher, "I Know I Also Share the Guilt," p. 257.

219 *"If Germany lifts"*: *Jerusalem Post,* March 2, 1965.

"There's going to be": Interview with Zeev Sharon.

220 *"Do you know who this is?"*: Interview with Amnon Meidad.

Twenty-Six: The Legislator

222 *"Mr. President"*: Official transcript of the Bundestag's debate, March 10, 1965. Unless otherwise noted, all of the quotes from the legislators in this chapter are from this source.

223 *"our duty is to ward off scandal"*: Neal Ascherson, "Bundestag Speaks Up," *Jerusalem Post,* March 16, 1965.

A few days earlier: New York Times, March 11, 1965.

"winter sun lay golden stripes": Ascherson, "Bundestag Speaks Up."

"He is an ungainly figure": Ibid.

"Future perpetrators [would] think": Buscher, "I Know I Also Share the Guilt," p. 286.

224 *"We all knew, really"*: Ascherson, "Bundestag Speaks Up."

225 *"He said something"*: Ibid.

WAS CUKURS TO BE KIDNAPPED: *Frankfurter Allgemeine Zeitung,* March 13, 1965.

THE CUKURS AFFAIR: *Neue Zürcher Zeitung,* March 10, 1965.

"The [Cukurs] case came to light": Associated Press, *Delaware County Daily Times,* March 19, 1965.

226 *"I do not know anything"*: William Buckley, "Lynch Justice Is Far

Better Than Tampering with Judicial Systems," *Valley Morning Star* (Harlingen, TX), March 18, 1965.

"All those years": "Tuviah Friedman, 1922–2011," TheWeek.com, February 10, 2011, https://theweek.com/articles/487360/tuviah-friedman-19222011.

227 *"The success of Wiesenthal's public struggle"*: Segev, *Simon Wiesenthal*, p. 192.

"a victory for the moral conscience": Ascherson, "Bundestag Speaks Up."

"The murder of an old Latvian": "The Murder of an Old Latvian Nazi," editorial, *Bergen (NJ) Record*, March 29, 1965.

228 *"To tell the truth"*: Yitzhak Arad, *The Operation Reinhard Death Camps: Belzec, Sobibor, Treblinka* (Bloomington: Indiana University Press, 2018), p. 186.

"[The] execution had a profound effect": Kuenzle and Shimron, *The Execution of the Hangman of Riga*, p. xxi.

229 *"No freedom for killers!"*: *New York Times*, July 4, 1979.

230 *"We cannot bow"*: Ibid.

"Today . . . ," Friedman wrote: Friedman, *The Struggle for the Cancellation of the Statute of Limitations*, p. 14.

"The photograph of Cukurs' corpse": Kuenzle and Shimron, *The Execution of the Hangman of Riga*, p. 141.

231 *"My father wasn't capable"*: Interview with Amnon Meidad.

"a general without troops": Interview with Gad Shimron.

"People were grateful": Interview with Amnon Meidad.

When the team gathered: Interview with Lihi Yariv-Laor.

"If they needed an old fox": Interview with Gad Shimron.

"He looks like a grandfather": "Dies ist Mein Mörder."

232 *"My father was very proud"*: Interview with Lihi Yariv-Laor.

"Cukurs snatched the baby": Press, *The Murder of the Jews in Latvia*, p. 159.

"While there is absolutely no doubt": Email interview with David Silberman.

233 *"a dashing guy"*: Von Lorenz Hemicker, "SS-Scherge wird Musical-Star," *Frankfurter Allgemeine Zeitung*, October 9, 2014.

"Hundreds of people": Cynthia Blank, "Latvian Nazi Musical Stirs

Jewish Outrage," Israel National News, October 31, 2014, http://www.israelnationalnews.com/News/News.aspx/186861.

Twenty-Seven: The Asylum

235 *In the last years:* USC Shoah Foundation, Visual History Archive, interview with Zelda-Rivka Hait. All of the quotes and details in this chapter about Nank and Zelma's experiences in Soviet-controlled Latvia are from this source.

242 *"an unimaginable chamber of horrors"*: Robert G. Kaiser, "Norilsk, Stalin's Siberian Hell, Thrives in Spite of Hideous Legacy," *Washington Post,* August 29, 2001.

244 *She wrote Nank:* Interview with Naomi Ahimeir. All of the quotes from Naomi in this chapter are from this source.

247 *"I would hear him mutter"*: Interview with Paul Semenoff.

Twenty-Eight: The Trial

249 *In Tel Aviv, Zelma:* USC Shoah Foundation, Visual History Archive, interview with Zelda-Rivka Hait. Unless otherwise noted, all of the quotes and details in this chapter about Zelma's experiences after immigrating to Israel are from this source.

"We talked every single day": Interview with Naomi Ahimeir. All of the quotes from Naomi in this chapter are from this source.

251 *"My second marriage"*: Silberman, *The Right to Live,* p. 74.

253 *"asking for refuge"*: Ezergailis, "Sonderkommando Arājs," p. 7.

"famous anti-Semite": Testimony of Ber Menkelkorn, Viktor Arājs trial, Hamburg State Archive.

"Stalin called me for a meeting": Leonid Leparyonok, "Prominent Russians: Alexsandr Yakovlev," Russiapedia, https://russiapedia.rt.com/prominent-russians/space-and-aviation/aleksandr-yakovlev/.

254 *Yakovlev called the flier to Moscow:* The details about the Yakovlev-Cukurs meeting are from Šāberte, *Laujiet man runāt!,* pp. 103–4.

"That day I was playing": Ibid., p. 104.

255 *"It did not take much"*: Ezergailis, "Sonderkommando Arājs," p. 6.

It seems almost certain: Interestingly, Cukurs may have gone to Arājs because the latter had a history with the Soviets himself. When he was younger, Arājs became interested in Marxism, even receiving a degree in Soviet law as a young man. "Indubitably," he said, "I was then a communist." Ibid.

accepted the aviator's offer: Ibid., p. 19.

256 *"The war of all wars":* Plavnieks, *Nazi Collaborators on Trial,* p. 76.

"the Nazis were able to convince": Ibid., p. 62.

"I knew": Ibid., p. 68.

257 *"Only because of the folly":* Vīksne, "Members of the Arājs Commando," p. 201.

"reptilian opportunist": Mihail Sebastian, *Journal, 1935–44* (Lanham, MD: Rowman and Littlefield, 2013), p. x.

258 *"I survived the terrible slaughter":* Silberman. *The Right to Live,* p. 74.

"long, affectionate, philososophical letters": Interview with Naomi Ahimeir.

PHOTO CREDITS

HERBERT CUKURS The National Archives of Latvia. The Latvian State Archive of Audiovisual Documents; **CUKURS WITH PLANE** Keystone-France/Getty Images; **RIGA CITIZENS** Bundesarchiv; **ZELMA SHEPSHELOVICH** Naomi Ahimeir; **YOUNG ABRAM SHAPIRO** Paul Semenoff; **JĀNIS VABULIS** Naomi Ahimeir; **MASS GRAVE** Bundesarchiv; **SIGN** United States Holocaust Memorial Museum, courtesy of Emmi Lowenstern; **TUVIAH FRIEDMAN** Mirrorpix; **SIMON WIESENTHAL** Courtesy of Simon Wiesenthal Center Archives; **MIO** Private Collection; **YOSEF YARIV** Lihi Yariv-Laor; **ELIEZER SUDIT AND WIFE** Zeev Sharon; **ADOLF ARNDT** Bundesarchiv

INDEX